Practical Electronics
A Self-Teaching Guide

Ralph

John Wiley & Sons, Inc.

This book is printed on acid-free paper. ∞

Copyright © 2004 by Ralph Morrison. All rights reserved

Published by John Wiley & Sons, Inc., Hoboken, New Jersey
Published simultaneously in Canada

No part of this publication may be reproduced, stored in a retrieval system, or transmitted in any form or by any means, electronic, mechanical, photocopying, recording, scanning, or otherwise, except as permitted under Section 107 or 108 of the 1976 United States Copyright Act, without either the prior written permission of the Publisher, or authorization through payment of the appropriate per-copy fee to the Copyright Clearance Center, 222 Rosewood Drive, Danvers, MA 01923, (978) 750-8400, fax (978) 750-4470, or on the web at www.copyright.com. Requests to the Publisher for permission should be addressed to the Permissions Department, John Wiley & Sons, Inc., 111 River Street, Hoboken, NJ 07030, (201) 748-6011, fax (201) 748-6008, email: permcoordinator@wiley.com.

Limit of Liability/Disclaimer of Warranty: While the publisher and the author have used their best efforts in preparing this book, they make no representations or warranties with respect to the accuracy or completeness of the contents of this book and specifically disclaim any implied warranties of merchantability or fitness for a particular purpose. No warranty may be created or extended by sales representatives or written sales materials. The advice and strategies contained herein may not be suitable for your situation. You should consult with a professional where appropriate. Neither the publisher nor the author shall be liable for any loss of profit or any other commercial damages, including but not limited to special, incidental, consequential, or other damages.

For general information about our other products and services, please contact our Customer Care Department within the United States at (800) 762-2974, outside the United States at (317) 572-3993 or fax (317) 572-4002.

Wiley also publishes its books in a variety of electronic formats. Some content that appears in print may not be available in electronic books. For more information about Wiley products, visit our web site at www.wiley.com.

Library of Congress Cataloging-in-Publication Data:

Morrison, Ralph.
    Practical electronics : a self-teaching guide / Ralph Morrison.
      p. cm.—(Wiley self-teaching guides)
    Includes bibliographical references and index.
    ISBN 0-471-26406-7 (pbk.)
      1. Electronics—Programmed instruction. I. Title. II. Series.

TK7816.M69 2003
621.381'07'7—dc21
                                                      2003050150

Printed in the United States of America

10 9 8 7 6 5 4 3 2 1

Contents

List of the Learning Circuits — v
A Note to the Reader — ix

1 Resistors, Capacitors, and Voltage — 1

2 Inductors, Transformers, and Resonance — 41

3 Introduction to Semiconductors — 81

4 More Semiconductor Circuits — 119

5 Feedback and IC Amplifiers — 151

6 IC Applications — 187

7 Circuit Construction, Radiation, and Interference — 217

8 A Review of Basic Electrical Concepts — 237

Appendix I: Preparing to Use the Learning Circuits — 261
Appendix II: Basic Algebra — 267
Index — 275

Learning Circuits

1 Resistors, Capacitors, and Voltage

    LEARNING CIRCUIT 1   Measuring Battery Voltages

    LEARNING CIRCUIT 2   Combining Resistors in Series and Parallel

    LEARNING CIRCUIT 3   Predicting and Measuring Voltage at the Junction of Three Resistors

    LEARNING CIRCUIT 4   Determining the Internal Impedance of a Battery

    LEARNING CIRCUIT 5   Observing Current Flow in a Capacitor

    LEARNING CIRCUIT 6   Observing the RC Time Constant

    LEARNING CIRCUIT 7   Constructing and Observing a Low-Pass RC Filter

    LEARNING CIRCUIT 8   Constructing and Observing a High-Pass Filter

2 Inductors, Transformers, and Resonance

    LEARNING CIRCUIT 9   Observing the Rise in Current for a Fixed Voltage in an Inductor

    LEARNING CIRCUIT 10   Observing the Response of an RLC Circuit

    LEARNING CIRCUIT 11   Observing the Response of a Parallel Resonant Circuit

    LEARNING CIRCUIT 12   Building a DC Power Supply

    LEARNING CIRCUIT 13   Building a Voltage Doubler

3 Introduction to Semiconductors

LEARNING CIRCUIT 14	Observing How a Diode Is Used to Clamp a Sine Wave
LEARNING CIRCUIT 15	Diode Clamping to a Power Supply
LEARNING CIRCUIT 16	Observing Voltage Regulation Using a Zener Diode
LEARNING CIRCUIT 17	Observing Voltage Clamping with Two Zener Diodes
LEARNING CIRCUIT 18	Constructing an Emitter Follower
LEARNING CIRCUIT 19	Observing the Output Impedance of an Emitter Follower
LEARNING CIRCUIT 20	Measuring the Gain of an Emitter Follower
LEARNING CIRCUIT 21	Obtaining Voltage Gain from a Transistor
LEARNING CIRCUIT 22	Obtaining Voltage Gain from a Transistor at AC
LEARNING CIRCUIT 23	Observing the Gain of a PNP Emitter Follower
LEARNING CIRCUIT 24	Providing Voltage Gain Using an NPN and a PNP Transistor

4 More Semiconductor Circuits

LEARNING CIRCUIT 25	The Stacked Emitter Follower and How It Functions
LEARNING CIRCUIT 26	Building a Positive Voltage Regulator
LEARNING CIRCUIT 27	Building a Negative Voltage Regulator
LEARNING CIRCUIT 28	Building Two Constant Current Sources
LEARNING CIRCUIT 29	Building a Differential Input Circuit
LEARNING CIRCUIT 30	Building a Transistor Switch
LEARNING CIRCUIT 31	Building a Switch Using a Transistor

5 Feedback and IC Amplifiers

- LEARNING CIRCUIT 32 — Mounting an IC Amplifier DIP Socket
- LEARNING CIRCUIT 33 — Constructing and Testing a Gain 1 IC Amplifier
- LEARNING CIRCUIT 34 — Using Feedback and an IC Amplifier to Provide Positive Gain
- LEARNING CIRCUIT 35 — Constructing and Testing a Negative Gain IC Voltage Amplifier
- LEARNING CIRCUIT 36 — Using Feedback to Correct for Signal Distortion
- LEARNING CIRCUIT 37 — Building and Applying a Differential Amplifier
- LEARNING CIRCUIT 38 — Building and Testing an Active Second-Order Low-Pass Filter
- LEARNING CIRCUIT 39 — Building and Testing a High-Pass Circuit

6 IC Applications

- LEARNING CIRCUIT 40 — Building and Testing a Regulated Power Supply
- LEARNING CIRCUIT 41 — Building and Testing a Sawtooth Voltage Generator
- LEARNING CIRCUIT 42 — Observing the Operation of a Bistable Multivibrator

A Note to the Reader

Welcome to the study of electronics. This is a fascinating field, one whose relevance to life in the twenty-first century cannot be overestimated. Our lives have been changed immeasurably by electronic devices from cell phones to computers. Electronics are used in medicine, in aviation and space travel, in the cars we drive, and in the many devices in our homes. Indeed, some form of electronics touches every aspect of our lives. Yet these countless devices are built from the same basic components. So the study of basic electronics will give you a way to understand how many of the electronics we use every day actually function.

My approach has been to start with the fundamentals and build up your knowledge step by step. In chapter 1 you'll begin working with just two of the simplest electronic components, and will learn how to combine them into circuits. Then you will add more components and move on to more complicated circuits.

Understanding electronics depends on your having some familiarity with electricity itself. You need to know the basics of voltage, current, and Ohm's law. If it has been some time since you studied basic electricity, I have provided a review for you in chapter 8. If you need to brush up on these concepts, chapter 8 is the place to start, even before you start on chapter 1.

If you are completely unfamiliar with electricity, you may find that the review in chapter 8 is not enough. In this case it makes sense to prepare yourself for the study of electronics by reading *Basic Electricity*, which is also part of this series of Self-Teaching Guides.

The basic principles of electronics are ultimately derived from physics. Many of these principles can be expressed most simply and completely as algebraic equations. In an equation, letters and symbols are used to represent the components and variables used in electronics. Algebra provides us with techniques for stating principles and solving problems using these letters and symbols. You will find a number of equations used in this book. If it has been some time since you used algebra, you will find a review of it in appendix II. If you are completely unfamiliar with algebra, there are many textbooks on elementary algebra available. An understanding of trigonometry will help in some sections of the book, but it is not required and you will be able to proceed without it.

One feature of this book that I think you will find enjoyable and useful is the Learning Circuits. A Learning Circuit is an electronic circuit you can build yourself, which will help you to understand a principle of electronics. In order to build the Learning Circuits you need a supply of basic components. These are not expensive and are readily available at electrical supply stores. Each Learning Circuit begins with a list of the components you need to build it, and has complete instructions and drawings to help you put it together. A complete list of all the components needed for every Learning Circuit appears in appendix I. Appendix I also has information about the testing equipment used in the Learning Circuits, and some general advice about putting together electronic circuits.

It is possible to go through the book and understand the material without building the Learning Circuit exercises, but you will get much more out of the book if you do. It's also fun. One of the most enjoyable parts about writing this book was building the Learning Circuits, which I did with a student to whom it was all brand-new. There is truly no substitute for hands-on experience, and this is a good way to acquire it at your own pace.

1 Resistors, Capacitors, and Voltage

Objectives

In this chapter you will learn:

- two definitions of electronics (and how to tell which one is intended)
- how to study electronics using the Learning Circuits
- the equipment you will need for the Learning Circuits
- the characteristics of two basic components used in most electronic equipment—the resistor and the capacitor

What Is Electronics?

The word *electronics* has two different, though closely related, meanings. This can be confusing, but you will find you can quite easily tell which meaning is intended. In the first definition, electronics is "the study of voltage and current waveforms that vary in time." When this meaning is intended, the word *electronics* is singular, and in a sentence it is used with a singular verb. For example, one would say, "Electronics *is* the study of voltage and current waveforms."

Electronics can also refer to electrical devices created to perform specific tasks, such as amplifying an electrical signal, sending or receiving radiation, or any one of hundreds of different functions. If this is the sense intended, the word *electronics* is plural and is followed by a plural verb. One would say something like, "The electronics onboard an airplane *are* very sophisticated."

You will be able to tell whether electronics refers to the study or to the devices by observing the way the word is used in a sentence and the context in which it is used. In ordinary conversation, the second sense of the word is used most often. If you walk into almost any large department store, you will find a section of the store called electronics. It's the section where you can buy DVDs, CD players, and so on. Clearly, the word refers to electronic devices. In this book, on the other hand, the word *electronics* refers most of the time to the study of voltage and current waveforms. That is why this section is called "What Is Electronics?" and not "What Are Electronics?"!

The Learning Circuits

Throughout this book you will find experiments in electronics you can do yourself. They are called *Learning Circuits,* and they have been designed to give you a hands-on sense of the way electronic circuits work. A circuit is a group of interconnected electronic components. They perform such tasks as amplification, waveform generation, filtering, signal sensing, signal switching, logic, radiation, and electromagnetic field detection.

Don't worry if you don't know what these functions are. By the time you finish this book you will be familiar with all of them. They are the functions that make up radios, VCRs, stereo amplifiers, telephones, and all the other electronic devices we use. You will not be able to design these electronic products when you finish this book (that requires more advanced study), but you will have a much better appreciation of how they work, and you will be well prepared to take the next step toward learning to design them yourself.

In this chapter there are 8 Learning Circuits. These first experiments will show you different ways of connecting two basic electrical components: resistors and capacitors. Before you create your first circuit, however, you need some equipment and some understanding of how to use it.

First, you need some means of measuring and observing changing waveforms. The Learning Circuits include pictures of changing voltage waveforms you can expect to see, and equations describing them. Pictures and equations are helpful, of course, but there is no substitute for seeing the voltages in real time, and for this you need some equipment. The two main tools used in electronics to make observations are the *waveform generator* (also called a *function generator* or signal generator) and the *oscilloscope*.

So, should you dive in and immediately purchase these two pieces of equipment? That will be your decision, but be sure to read appendix I, "Preparing to Use the Learning Circuits," first. As you'll see, the equipment is costly. Before making the purchases, you might want to go through at least a chapter or two of this book, studying the Learning Circuits and the drawings that accompany them. You can certainly learn a great deal this way. Then, if you find you are still excited about electronics, you can look for some used equipment and start making your own observations.

For the first few Learning Circuits, you can use another piece of equipment called a *multimeter*. As the name implies, this is a multiuse measuring device that can function as a voltmeter, ohmmeter, or ammeter. Multimeters are not very expensive, and they can measure ac and dc volts, dc current, and resistance. What they cannot do is show waveforms—for that an oscilloscope is needed.

Circuits also need a source of power, but using utility power from a wall plug poses a safety hazard. To resolve this difficulty, all of the ac sources in the book make use of an ac *adapter*. An adapter is an Underwriter's approved transformer that supplies a source of low-voltage ac power. (Underwriter's Laboratory is a testing organization that approves electrical hardware for use by consumers.)

Connecting circuit components together requires some tinned bus wire, some insulated wire, some solder, and a soldering iron. Simple circuits can be connected using clip leads or test leads. A test lead is an insulated wire that has mechanical clips (alligator clips) on each end. In some circuits you can make a connection simply by twisting leads together. You can also purchase a circuit board that has a grid of holes so that tie pins can be pressed into the holes, and you can solder components to these pins. (Just be sure not to cut the leads short until you are certain they are resting in their final location.) But when circuits become more complicated, soldering works best. You will find a step-by-step description of soldering in appendix 1.

You will also need some basic tools, which you may already own: a pair of needle-nosed pliers and wire cutters.

Finally, you will need a workstation to do the Learning Circuits, which does not need to be more than a few feet of counter space near a wall outlet. Ideally this should be a place where you can leave your equipment out and available while you experiment with the various Learning Circuits.

The Waveform Generator

A waveform (function) generator is a piece of electronic test equipment used to generate a repetitive changing voltage, or a voltage waveform that repeats itself over and over. The voltage waveforms that can be selected are sine, square, or triangle. The lowest settable frequency is often around 0.1 Hz (see chapter 8). The highest sine wave frequency is often 10 MHz. The voltage amplitude is often limited to 10 V peak or 20 V peak-to peak.

Each of these waveforms has its own particular use. Sine waves are used to test the response of circuits. A sine wave is sinusoidal in character. A sine wave is often referred to as a sinusoid. See Figure 1.1. This waveform is used because the currents and voltages in a linear circuit are all sine waves. Square waves are valuable because they provide information about circuit behavior not easily seen with sine waves. A square wave voltage can transition symmetrically around 0 volts or transition from 0 to a peak voltage level once per cycle. The transition time or rise time should be short compared to the time of one cycle. This makes it difficult to generate a 2-MHz square wave, as the transition times should be around 5.0 ns. Shorter transition times raise the cost. Triangular waves are useful because the voltage slopes are constant. However, triangular waves are not generally used in testing. These three waveforms are shown in Figure 1.1. For more on these concepts, see chapter 8.

An output cable is usually supplied with the waveform generator (which can also be called a *signal generator* or a *function generator*). The cable can have alligator clips on the end so that it can be connected to various points in a circuit. The outer conductor of the cable is called the

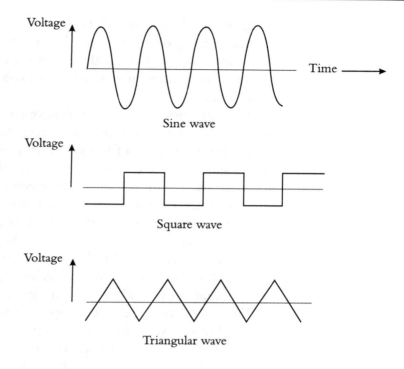

Figure 1.1 The three voltage waveforms provided by a waveform generator

zero reference conductor of the signal. It is also called a *shield,* a *ground,* or the *common* conductor. (The word *ground* is often used to mean "earth" but this is not always the intended meaning.) It connects to the ground or common of the circuit you are testing. In some generators this shield is connected to the safety or green wire of the power conductor. For most testing, you should remove this connection link or strap. The circuit you are testing may already be connected to ground. If this is the case, then two connections to ground can be troublesome. This can be checked by measuring a low resistance from the common output lead to the third pin on the power cord.

The Oscilloscope

An oscilloscope is a piece of electronic test equipment used to observe circuit behavior in real time. The oscilloscope generates a picture of the changing signal patterns. All of the waveforms shown in the figures in this book can be observed in real life by applying a signal generator to a circuit and then observing the waveforms with an oscilloscope. The vertical scale on an oscilloscope displays voltage, and the horizontal axis displays time.

The operation of a basic oscilloscope is simple. A dot moves (transitions) linearly across a viewing screen from left to right. When it reaches the right edge of the screen it immediately returns to the left side. A single crossing is called a "sweep." If the sweep frequency is set to 1 kHz, the dot moves across the screen in 1 millisecond (ms). In the first sweep, the time goes from 0 to 1 ms. The dot returns to the start and traces the same path for the second millisecond, and so on.

If the voltage probe is connected to a 1-kHz sine wave voltage, a single sine wave will be displayed on the screen. If the sine wave frequency is 2 kHz, then two full sine waves will be displayed. When the dot makes many sweeps per second, the screen pattern appears stationary.

You can observe the oscilloscope display in slow motion by observing a 1-Hz sine wave from a function generator, with the sweep frequency on the oscilloscope set to 1 Hz. At this slow rate you will be able to see the dot move across the screen, writing a sine wave pattern over and over. A sine wave voltage display is shown in Figure 1.2.

It is worthwhile spending a little time with the oscilloscope and the function generator before you get started on the Learning Circuits. I can't tell you exactly how to work the controls, as there are many different designs. You will have to hook up the oscilloscope to the function generator and play with the dials until it becomes clear. Don't worry, you can't hurt yourself or the equipment, so go ahead and experiment.

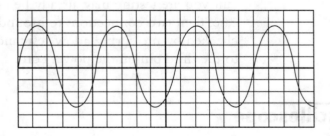

Figure 1.2 A sine wave displayed on an oscilloscope

An oscilloscope has one or two input probes. The probe tip is designed to connect to points in a circuit. The grounding clip on the probe is usually connected to the zero reference or ground of the circuit. At its other end (inside the oscilloscope) the grounding clip connects to the oscilloscope frame and to the power safety or green wire (the third plug in a three-pronged electrical plug). This connection to ground is required by the National Electrical Code.

A problem arises when you have both an oscilloscope and a function generator (or in more complicated circuits, multiple devices) all connected to one circuit. If two or more devices are connected to the power safety, you have multiple grounds. This is not desirable. For this reason, when you do have several devices connected to your circuit, use a "cheater plug" (a two-pronged plug) for all but one of your devices. In the Learning Circuits you will be using an adapter plug, which provides an additional level of isolation and safety.

Voltages

Throughout this book, the figures generally include a reference to a voltage source. The voltage source can be either ac or dc (see chapter 8). Voltage may come from a waveform generator, a battery, or the power utility.

The symbol used for a voltage source is either the letter V in a circle or the symbol for a battery. A lowercase v refers to a changing voltage. The polarity (plus or minus) of a dc voltage will always be indicated.

We will assume that the voltage source can supply the current demanded by the circuit without changing voltage. This is referred to as an ideal voltage source. In actuality voltage does change with load, but assuming an ideal source simplifies the discussion.

If the voltage is a step function or square wave, it will be clearly stated in the text.

Resistors and Capacitors

We are now ready to begin using our first electronic components, resistors and capacitors. These two components are found in most electronic equipment because they do very basic and important jobs needed in every circuit.

Resistors are the most common electrical component (see chapter 8). They are used to limit the flow of current in a circuit. By comparison a conductor offers very little opposition to current flow. There are many types and sizes of resistors. In electronics, resistors are apt to be small cylinders that are about a half-inch long. This is the circuit symbol for a resistor:

Capacitors are the second most common component. Their basic function is to store electrical field energy. This field energy requires electric charge on the plates of the capacitor. (See chapter 8 for discussion of capacitors and electric charge.) The ratio of voltage to charge is called *capacitance*. Since it takes time to store energy, capacitors can be used to control frequency response, provide filtering action, provide timing, and store energy in power supplies. Capacitors are found in almost every circuit design. The circuit symbol for a capacitor is

In the next sections we will be examining the way resistors and capacitors respond to various voltage waveforms. You will recall that a

Resistors, Capacitors, and Voltage   9

waveform generator produces sine waves, square waves, and triangle waves. Sine wave voltages are the only waveform that keep the same shape in any combination of resistors and capacitors.

One way to study resistors and capacitors is to apply a "step function" to the circuit. A step function is a voltage that changes from one value to another. In many cases a low-frequency square wave can be used as a step function. Digital circuits make extensive use of square waves and step functions.

A common source of dc voltage is the battery. Batteries can be placed in series to increase the dc voltage. If two 9-V batteries are placed in series, the total voltage is 18 V. This series arrangement is shown in Figure 1.3.

If one of the batteries is reversed in polarity, the voltages subtract.

In practice, series batteries should be of the same type so that the batteries will last the same length of time.

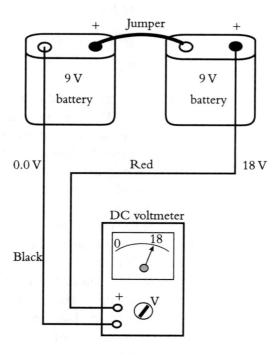

Figure 1.3 Batteries in series

> ⌇ **LEARNING CIRCUIT 1** ⌇
> ## Measuring Battery Voltages
>
> You will need (in addition to your multimeter or oscilloscope):
>   2 9-V batteries
> 1. Set the multimeter to "volts" and to a scale appropriate (the voltage can be easily read on that range) for measuring 9 to 18 V (usually 25 V). Practice using the multimeter (which is now functioning as a voltmeter) by measuring the voltage of a battery. It should read 9 V.
> 2. Touch the batteries together in series, positive to negative, so the voltages add. Measure the total voltage. The reading should be 18 V.
> 3. Tie the common voltmeter lead to the connection between the two batteries. Note that the ends of the batteries are at +9 V and −9 V.
> 4. Do the same measurements using the oscilloscope. Note the trace moves across at different levels.

When batteries are placed in series, any one of the battery terminals can be called 0 V. If the midpoint or jumper between the two series 9-V batteries is called 0 V, then the other two terminals are at −9 V and +9 V. The potential difference between the outer terminals is 18 V.

Batteries may be placed in parallel provided their voltages are equal. Connect the plus terminals together and connect the minus terminals together. The resulting voltage is the same as one of the batteries, but the current capability is increased. A problem with this arrangement is that if one of the batteries becomes weak, the strong battery will drain into the weak battery.

AC voltage sources can be placed in series if they are voltages on separate windings or coils of a transformer. (Refer to chapter 8.) The voltages must be in phase (peak at the same time) if the voltages are to add. The coils of the transformer should have nearly the same dc resistance for this to be practical. In utility power generation, ac sources (generators) are placed in parallel on the power grid. Placing a power generator on line requires skill and a complex procedure.

Resistors, Capacitors, and Voltage

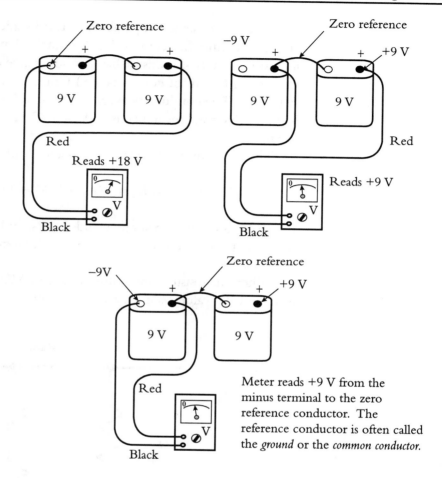

Figure 1.4 Two 9-V batteries and a reference conductor

Resistors in Series or Parallel

When resistors are placed in series, their resistances add. If three 10-ohm resistors are placed in series, the total series resistance is 30 ohms. To calculate the total resistance, the units of resistance must agree. From this point on we will use the Greek letter Ω to mean ohm.

Consider a 300-Ω resistor in series with a 2-kΩ resistor. The 200 Ω must be expressed as kΩ or the 2 kΩ must be expressed as ohms. The answer is 2 kΩ + 0.3 kΩ = 2.3 kΩ. The other solution is 2,000 Ω + 300 Ω = 2,300 Ω.

The conductance of a resistor is the reciprocal of its resistance. You calculate the resistance of parallel resistors by adding their conductances, and then taking the reciprocal of the total. Figure 1.5 shows two parallel resistors. A 10-Ω resistor has a conductance of 0.1 S, where S stands for sieman (a unit of conductance.) Consider a 5-Ω and a 2-Ω resistor in parallel. The conductances are 0.2 S and 0.5 S. The total conductance is 0.7 S. The total resistance is ⅟₀.₇ Ω = 1.429 Ω. This circuit is shown in Figure 1.6.

The equation relating the conductances of three parallel resistors is

$$1/R_T = 1/R_1 + 1/R_2 + 1/R_3 \quad (1.1)$$

where R_T is the total resistance and R_1, R_2, and R_3 are the resistances of the three parallel resistors. This idea can be extended to any number of resistors.

When the resistors are in units of kΩ or MΩ, a convenient technique is to treat the values as if they were all ohms and at the end of the calcu-

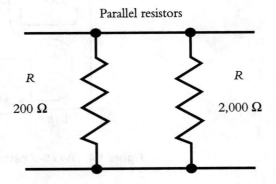

The letter representing a resistor is usually a capital R.

The unit of resistance is the ohm, abbreviated by the Greek letter Ω.

200 ohms can be written 200 Ω.

200 ohms can be written 0.2 kΩ where k stands for 1,000.

The parallel resistance of these two resistors is 181.81 Ω.

Figure 1.5 Resistors in parallel

Parallel resistors and a voltage source

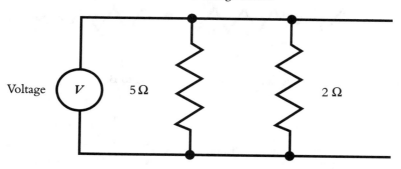

The 5-Ω resistor has a conductance of 0.2 sieman. The abbreviation for sieman is S. The 2-Ω resistor has a conductance of 0.5 S. The conductances add together and equal 0.7 S. The resistance is the reciprocal of the conductance, or 1.429 Ω. The voltage V can be from a battery or a signal from a generator.

Figure 1.6 A circuit with two parallel resistors

lation affix the correct unit. (The capital M stands for million. It is pronounced mega. A lowercase m stands for thousandth. It is pronounced milli. A lowercase k stands for thousand. It is pronounced kilo.) For example, 10 MΩ in parallel with 20 MΩ is treated like 10 and 20 Ω. The answer is 6.66 Ω. With the correct units the answer is 6.66 MΩ.

LEARNING CIRCUIT 2
Combining Resistors in Series and Parallel

You will need (in addition to your multimeter or oscilloscope):

3 1,000-Ω resistors

Set your multimeter to ohms × 1,000 (it is now functioning as an ohmmeter). Measure the resistances of two 1,000-Ω resistors in series and parallel combinations. Do the same thing using three resistors. How many combinations of series and parallel can you make with three resistors? See Figure 1.7.

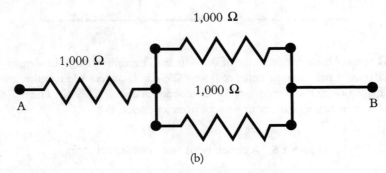

(a)

The resistors in series provides a 3,000-Ω resistor from A to B.

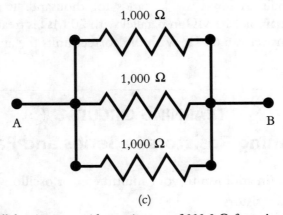

(b)

This arrangement provides a resistance of 1,500 Ω from A to B.

(c)

These parallel resistors provide a resistance of 333.3 Ω from A to B.

Figure 1.7 The three ways three 1,000-Ω resistors can be arranged

Voltages Applied to Resistors

When a voltage is placed across a resistor, a current flows in a loop formed by the voltage source and the resistor. This circuit is shown in Figure 1.8.

The current in the loop is given by Ohm's law. By convention, the direction of the current is out of the positive terminal of the battery. The current level for a 6-V battery and a 1,000-Ω resistor is 6 mA. The power dissipated in the resistor is given by V^2/R. (See chapter 8 for discussion of these concepts.) To use this equation, the voltage must be expressed in volts and the resistor in ohms. The power is 36/1,000 W or 36 mW. If the voltage were 6 V ac, the answer would be the same.

Standard carbon resistors are commercially available that cover the range 10 Ω to 22 MΩ. These resistors are available in ¼-W, ½-W, 1-W, and 2-W sizes. It is good practice to avoid using resistors at more than one-half their wattage rating. In most circuit applications it is convenient to use resistors of one wattage size, as ½-W size is a typical power level. This means many resistors are rated higher than they need to be. The standard resistor values, with accuracy to within 20% of the stated value, in the range from 10 to 100 Ω are 10, 12, 15, 18, 22, 27, 33, 39, 47, 56, 68, and 82 Ω. These same multiples are available in every decade. For example, resistors are available at 22, 220, 2.2 k, 22 k, 220 k, 2.2 M and 22 MΩ. Resistors with accuracy to within 10% and 5% of their stated value are also available.

The value of a carbon resistor is noted in a rather esoteric way, by bands of color that encircle each end of the resistor. At one end of the

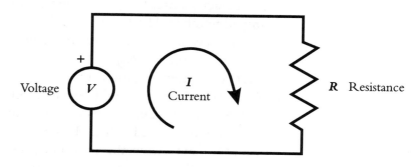

Ohm's law requires that $I = V/R$

Figure 1.8 A voltage source across a resistor

resistor, only silver and gold are used. Gold indicates that the resistor's accuracy is to within 5% of the stated value and silver indicates 10%. No color on this end indicates accuracy to within 20%.

On the other end of the resistor, 10 different colors plus gold and silver are used, in three bands. The first two bands indicate numeric values and the third band indicates the number of added zeros. The numeric meaning of each color is listed in the following table. Note that gold and silver on this end have different meanings than they have on the other end.

Black	0	Orange	3	Blue	6	White		9
Brown	1	Yellow	4	Violet	7	Gold (divide by 10)		
Red	2	Green	5	Grey	8	Silver (divide by 100)		

A resistor with bands of brown, black, and brown reads 100 Ω. The first brown is 1, the black band is 0, and the second brown indicates one added 0. The resistor with bands of brown, green, and green indicates 1,500,000 Ω. The brown is 1, the first green is 5, and the last green says that you add five zeros. Bands of green, blue, and gold indicate 5.6 ohms. Green is 5, blue is 6, and the gold says you divide by 10.

Another way resistors are made is by depositing a thin film of metal on a cylinder. A spiral is then cut into the surface of the cylinder. This cut defines the resistance—the closer together the spiral, the more resistance. These resistors are accurate to within 1% or 2%, and are more stable then carbon resistors.

For critical applications, precision wire-wound resistors are available with accuracies better than 0.1%. Metal film resistors are often coded by a stamped number. The last digit indicates the number of zeros.

If you have any doubt as to the value of a resistor, use your ohmmeter to verify the value.

Note that when you use your fingers to clamp the leads of the ohmmeter to the resistors, your fingers become part of the circuit. The resistance between your fingers can be as low as 10,000 Ω. The resistance depends on the individual, the surface area, the finger pressure, and the moisture and oils in the skin. For resistors over 1,000 Ω hold the ohmmeter by the plastic handles, and keep your fingers out of the circuit. When measuring a resistor, your fingers should not provide a parallel path for current flow, or the answer will be incorrect.

In many mass-produced products such as TV sets, the resistors are extremely small, have no leads, and are soldered directly onto the circuit board. At the other end of the spectrum, there are very large resistors that can dissipate 5, 10, or even 100 W. These resistors are made using a resistive alloy wire wound on a ceramic core.

The current that flows in an ideal resistor depends only on the instantaneous voltage. If the voltage is a sine wave, a step function, or a square wave, the current waveform is exactly the same. When the resistor value is very large or very small, there are exceptions to this statement. Resistors with low resistance values are influenced by their series inductance, and resistors with high resistance values are influenced by a shunt parasitic capacitance. These effects are important at high frequencies. For the circuits we will discuss in this book, these effects can be ignored.

The Voltage Divider

Two resistors in series across a voltage source form a voltage divider. This voltage divider circuit is shown in Figure 1.9.

One side of the voltage source is usually called the *reference conductor* or *zero of potential*. The voltage at the junction between the two resistors is a fraction of the total voltage. The purpose of the voltage divider is to produce this reduced potential. The ratio of voltage drops is equal to the ratio of resistance values. The sum of the two voltage drops is equal to the source voltage. In Figure 1.9 the negative side of the voltage source is the reference conductor. If the resistors are equal, the attenuation factor is 2. The current in this circuit is the voltage divided by the sum of the resistors, or $I = V/(R_1 + R_2)$. The voltage across the first resistor is this resistance times the current, or $V_1 = I \times R_1 = V \times R_1/(R_1 + R_2)$. The voltage across the second resistor is given by $V_2 = V \times R_2/(R_1 + R_2)$. For example, if $R_1 = 2$ kΩ and $R_2 = 8$ kΩ, their sum is 10 kΩ. If the voltage is 10 V, the current is 1 mA. The voltage across the 2-kΩ resistor is $I \times R_2$, or 2 V. If the negative terminal of the battery is at 0 V, then the connection between the two resistors is +2 V. If the positive terminal of the battery is at 0 V, then the junction between the resistors is at -8 V.

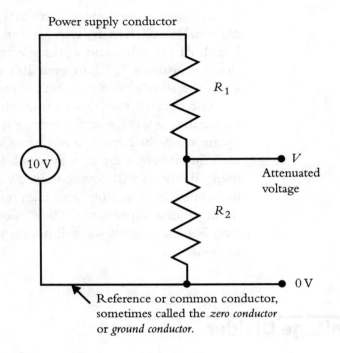

If $R_1 = R_2$, then the voltage V is 5 V. The voltage V is always less than 10 V. The two resistors are often called a *voltage divider*.

Figure 1.9 A voltage divider

For sine wave voltages or dc voltages, the power dissipated in each resistor can be calculated three different ways (see chapter 8). The voltage times the current, or $V \times I$, is the simplest way. The other two ways are I^2R and V^2/R. The units must be in ohms, amperes, and volts to get an answer in watts. Using the equation $P = V \times I$, the power dissipated in R_1 is $2 \times 0.001 = 2$ mW.

Sometimes it is necessary to select two resistors to obtain a required attenuation factor. The easiest way to solve this problem is to decide on a current level. The resistor values are simply the voltages divided by the current. For example, if 20 V is to be attenuated to 5 V, select a current of 1 mA. The resistor with 5 V is $5/0.001 = 5$ kΩ. The other resistor has 15 V across its terminals, so the resistor is $15/0.001 = 15$ kΩ.

> **LEARNING CIRCUIT 3**
>
> ## Predicting and Measuring Voltage at the Junction of Three Resistors
>
> You will need (in addition to your multimeter or oscilloscope):
>
> 3 1,000-Ω resistors
>
> 1 9-V battery
>
> 1. Place three 1,000-Ω resistors in series across a 9-V battery.
> 2. Place the black or negative lead of the voltmeter to the negative terminal of the battery.
> 3. Predict and measure the voltages at the junctions of the three resistors.
> 4. Measure the voltages placing the black voltmeter lead on the positive battery terminal. These are negative voltages if the positive battery terminal is used as the reference or zero conductor.

Source Resistance at DC

When current is taken from a practical voltage source, the voltage drops. This is the result of current flowing in an internal resistance. An internal resistor and an external load resistor form a voltage divider (attenuator). The circuit is shown is Figure 1.10.

Internal resistance of any circuit at the point of interest is the ratio of voltage change to current change.

One approach to determining the internal resistance (also called the *internal impedance*) of a circuit is to add a load resistor and note the voltage change. The voltage before the load is applied is called the *open circuit voltage*. The change in voltage, or voltage difference, is the open circuit voltage minus the voltage when current is flowing. The change in current is the loaded voltage divided by the load resistance. The ratio of these two numbers is the internal resistance.

For example, assume the open circuit utility power is 120 V at 60 Hz. A 12-Ω load resistor drops the voltage to 119 V. The voltage difference is 1.0 V. The current is 119/12 = 9.917 A. The internal resistance

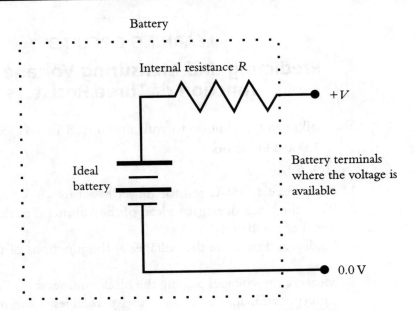

Source resistance is the value of R measured from the output terminals. The voltage V drops depending on the amount of current supplied by the battery. The voltage drop equals IR. The battery load and the internal resistance form an attenuator.

Figure 1.10 The internal resistance of a dc voltage source (also called source resistance)

is $1/9.917 = 0.1008\ \Omega$. The internal resistance of an ideal voltage source is zero. (*Warning:* Do not attempt to make this measurement on an open power connection. There is a good chance of electrical shock.)

It is sometimes useful to measure source resistance by using two different load resistors. For example, if the current varies from 15 to 35 mA and the output voltage varies from 9 to 8.98 V, the source impedance is $0.02\ \text{V}/0.02\ \text{A} = 1\ \Omega$.

The idea of source resistance can be extended to include inductance and capacitance. The term *source resistance* then becomes *source impedance*. The expression *output impedance* means the same thing. Impedance only has meaning for sine wave voltages and currents. The expression "*output impedance*" is in common use, but the value usually refers to a resistance.

The voltage divider in Figure 1.9 has an equivalent source resistance (series resistance) that can be determined by the ratio of open circuit

Resistors, Capacitors, and Voltage

> **LEARNING CIRCUIT 4**
>
> ### Determining the Internal Impedance of a Battery
>
> You will need (in addition to your multimeter or oscilloscope):
>
> 3 1,000-Ω resistors
>
> 1 9-V battery
>
> Use the 9-V battery and the three 1,000-Ω resistors in parallel. Place a voltmeter across the 9-V battery. Can you see the voltage drop when the resistor is touched to the battery terminals? Calculate the source resistance.

voltage to short circuit current. This is known as Thevenin's theorem. The open circuit voltage is $V_O = VR_1/(R_1 + R_2)$. The short circuit current is $I_{SC} = V/R_2$. The ratio of $V_O/I_{SC} = R_1R_2/(R_1 + R_2)$. This is the resistance of the two resistors in parallel, or the source resistance. In order to lower the source resistance, more current must flow in the divider. If both resistors are one-half of their first value, the source resistance would be one-half of its previous value. For example, consider a 20-V source where R_1 and R_2 are both = 10 kΩ. The equivalent circuit is a 10-V source in series with 5 kΩ. If R_1 and R_2 are each 5 kΩ, the source impedance is 2.5 kΩ.

Thevenin's theorem can be very impractical. A short circuit on the utility power line to measure source impedance would blow the breaker.

The Current Divider

When a voltage is applied across two parallel resistors as in Figure 1.11, the current in each resistor follows Ohm's law.

The total current can be calculated two ways. The first way is find the current in each resistor and then add the values together. The second way is to compute the parallel resistance and then calculate the current. If the resistors are 2.5 kΩ and 5 kΩ and the voltage is 25 V, the currents are 10 mA and 5 mA. The total current is 15 mA. Using the second

22 PRACTICAL ELECTRONICS

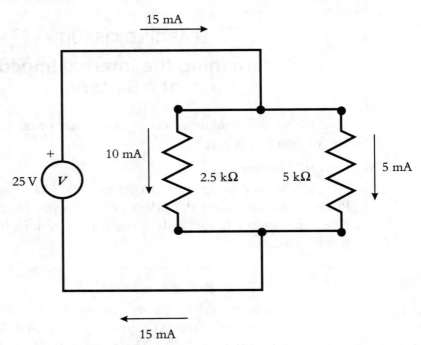

Figure 1.11 Parallel resistors across a voltage source

The current splits so that a portion flows in each resistor. The current flowing in each resistor is given by Ohm's law. If there were three resistors, the current would split three ways.

method, the resistors in parallel are 1.66 kΩ. The current is 25/1,666 = 0.015 A or 15 mA.

Switches

Switches provide a mechanical means of making or breaking one or more electrical connections. The moving part of a switch is called a *pole*. A pole can make one or more connections in the throw of a lever or the rotation of a shaft. The abbreviation for pole is the letter *p*. If the pole makes one connection when the switch lever is thrown, the switch is called *single-pole single throw,* or SPST. A pole that transfers between two connections is called an SPDT, or *single-pole double throw.*

Switches are available with several poles, with connected or disconnected center positions and with spring returns. Size, style, current rat-

RESISTORS, Capacitors, and Voltage

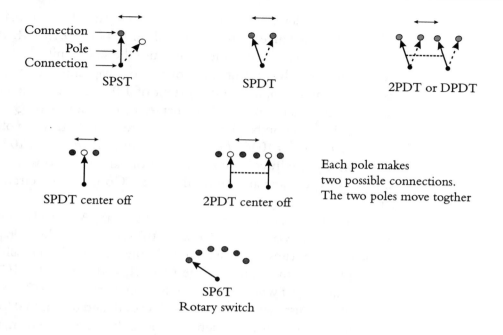

Momentary switches have spring returns so they cannot be left on or off.
Switches can be operated by push button, bat handle, lever, or knob.
There are toggle switches, sliding switches, and rotary switches. There are many different ways to draw a switch on a diagram. The diagrams presented here are typical.

Figure 1.12 Switch arrangements and their diagrams

ing, and mounting arrangements are also variables. There are thousands of combinations available for the electronics designer. A few common switch arrangements are shown in Figure 1.12.

Capacitors

When capacitors are placed in parallel, the capacitances add. Capacitances must be expressed in the same units for addition. For example, a 0.001 µF capacitor is in parallel with 500 pF. The problem requires that the 500 pF capacitance be expressed in µF. This is 0.0005 µF. The total capacitance is 0.0015 µF.

Capacitors in series add like parallel resistors. The sum of the reciprocal capacitances is equal to the total reciprocal capacitance. For example, consider a 0.2 μF in series with 0.4 μF. 1/0.2 + 1/0.4 = 3/0.4. The total capacitance is the reciprocal, or 0.4/3 = 0.133 μF. Again, to do a calculation all capacitances must use the same unit.

When a steady current (dc) flows into a capacitor, the voltage rises linearly. As an example, if a dc current of 1.0 mA flows into a capacitor for 1 ms, the charge is $I \times t$ (see chapter 8, Equation 8.4). $Q = 0.001 \times 0.001 = 10^{-6}$ coulombs. This much charge on 0.1 μF is a voltage $V = Q/C = 10^{-6}/10^{-7} = 10$ V. The voltage rises linearly from 0 to 10 V in 1 ms.

A steady rise in voltage in a capacitor can be demonstrated using a square wave voltage and an oscilloscope. Consider the circuit in Figure 1.13.

The square wave voltage is set to 10 V peak. A 100-kΩ resistor connects the square wave voltage to a 0.1 μF capacitor. If the voltage on the capacitor never rises to more than 0.1 V, the current is essentially constant at 0.1 mA. The maximum charge Q is $CV = 10^{-7} \times 0.01 = 10^{-9}$ C. This charge equals $I \times t$ where I is 10^{-4} A. Solving for t yields 10^{-5} seconds. The square wave must stay positive for 10^{-5} seconds and return to 0 for another 10^{-5} seconds. This is a frequency of 50 kHz. The waveform across the capacitor will be a triangle wave, a voltage that rises and falls in a linear manner. This voltage can easily be seen on the screen of the oscilloscope.

LEARNING CIRCUIT 5
Observing Current Flow in a Capacitor

You will need (in addition to your oscilloscope and function generator):

1 100-kΩ and 1 50-kΩ resistor (use two parallel 100-kΩ resistors)

2 0.1-μF capacitors

Connect a 0.1-μF capacitor and a 100-kΩ resistor to a square wave generator as shown in Figure 1.13. Set the frequency to 50 kHz and the amplitude to 10 V. Use the oscilloscope to observe and note the triangle wave across the capacitor.

Reduce the resistor to 50 kΩ and note that the amplitude of the triangle wave doubles. Now double the capacitance and note that the voltage returns to the first level.

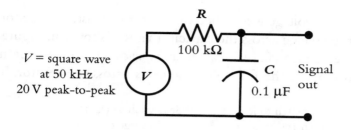

Circuit to supply current to a capacitor

V = square wave at 50 kHz
20 V peak-to-peak

R = 100 kΩ
C = 0.1 μF
Signal out

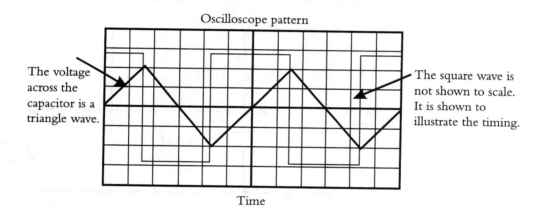

Oscilloscope pattern

The voltage across the capacitor is a triangle wave.

The square wave is not shown to scale. It is shown to illustrate the timing.

Time

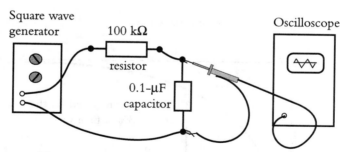

Square wave generator — 100 kΩ resistor — 0.1-μF capacitor — Oscilloscope

The wiring to the capacitor and the resistor

The current supplied to the capacitor is constant in both directions. During the time the current flows in one direction, the voltage changes linearly.

Figure 1.13 The voltage across the terminals of a capacitor when supplied a constant current for short periods of time

25

The RC Time Constant

When a dc voltage is switched on to a series resistor–capacitor (RC) circuit, the voltage across the capacitor rises as shown in Figure 1.14.

The voltage curve is called an *exponential curve*. At the moment of switch closure, the full voltage appears across the resistor. This means

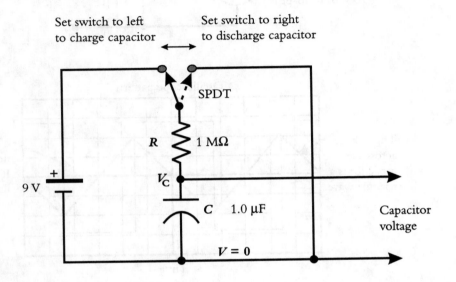

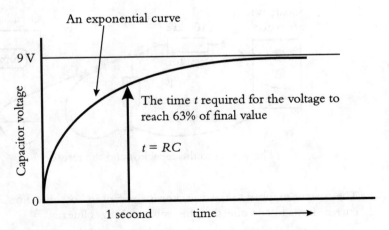

The rise in capacitor voltage can be observed on a voltmeter.

Figure 1.14 A circuit showing the rise of voltage in a series RC circuit

> ## ⊸ LEARNING CIRCUIT 6 ⊸
> ## Observing the RC Time Constant
>
> You will need (in addition to your multimeter or oscilloscope):
> 1 9-V battery
> 1 1-MΩ resistor
> 1 1-μF capicitor
> 1 SPDT switch
>
> Construct the circuit shown in Figure 1.15. Select $C = 1$ μF and $R = 1$ MΩ. The value of the time constant is $10^{-6} \times 10^{6} = 1$ second. This rise in voltage is slow enough that it can be observed on a voltmeter. If a square wave generator is used as the voltage source, the rise time can easily be seen. Set the sweep frequency on the oscilloscope to 1 Hz. Set the square wave frequency to 1 Hz. Observe the time it takes the voltage to reach 7 V.
>
> When a capacitor is discharged through a resistor, the falling voltage also follows an exponential curve. In this case the voltage falls to 37% in 1 time constant or 1 second. You can see this on the second half of the square wave cycle.

the initial current flow is V/R. As the capacitor receives charge, the voltage across the capacitor increases, taking away voltage from the resistor. At any moment in time the voltage across the resistor plus the voltage across the capacitor must equal the impressed voltage V. After a period of time, most of the voltage appears across the capacitor and there is very little current flow. The time it takes to reach 63% of final value is given by the product RC where R is in ohms and C is in farads. This is called the RC time constant.

To show that RC has units of time, we can use the following definitions. $R = V/I =$ volts/amperes and $C = Q/V =$ coulombs/volts. But coulombs = amperes × time. Therefore $RC =$ (volts/amperes) × (amperes/volts) × time. As you can see, the volts and amperes cancel, leaving the unit of time.

28 PRACTICAL ELECTRONICS

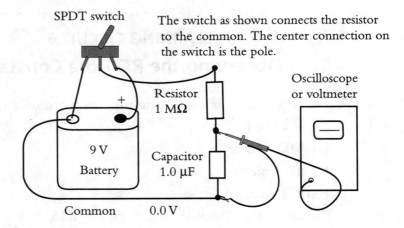

Figure 1.15 The construction of the circuit in Figure 1.14

The idea of a time constant can be applied a second or even a third time to the same circuit. In the previous example the voltage falls to 37% in 1 time constant (10 seconds). The voltage will fall to 37% of 37% in 2 time constants (20 seconds). This is a value of about 13%. In 3 time constants or 30 seconds the value is about 5%.

The Impedance of a Resistor and a Capacitor in Series

When a sinusoidal voltage is applied to a series RC circuit (shown in Figure 1.17), the current that flows is sinusoidal. The voltage across the resistor peaks at the same time the current peaks. The voltage across the capacitor lags the current by 90°. Since the peaks of voltage do not occur at the same time, the voltages cannot simply add together. In Figure 1.16 the three voltages are represented by rotating pointers (see chapter 8).

Each pointer makes a counterclockwise rotation once per cycle. The height of the pointers above the horizontal axis represents the instantaneous voltages. The length of the pointer is the peak value of voltage. When the voltage across the resistor is maximum, the voltage

Resistors, Capacitors, and Voltage 29

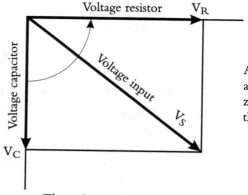

At time = 0 the voltage across the resistor is zero and the voltage across the capacitor is maximum.

The pointers rotate once per cycle.

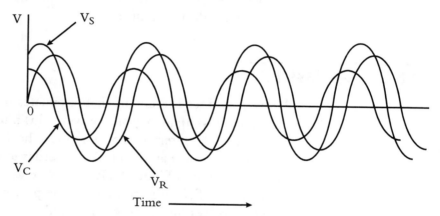

The three sine wave voltages displayed in time.

Figure 1.16 The rotating pointers for a series RC circuit

across the capacitor is 0. When the pointer for the resistor voltage points straight up, the pointer for the capacitor voltage points to the right.

To solve for the impedance in this circuit, it is convenient to assume a current I. The peak voltage across the resistor is IR. The peak voltage across the capacitor is IX_C where X_C is the reactance of the capacitor (see chapter 8). To find the total voltage V applied to the circuit, form a rectangle using the voltages across the resistor and capacitor as sides. The peak voltage V is the length of the diagonal of this rectangle. The length of this diagonal is the input voltage $V = \sqrt{(IX_C^2 + IR^2)}$. The ratio of peak voltage to peak current is

$$V/I = Z_{RC} = \sqrt{(X_C^2 + R^2)} \qquad (1.2)$$

This ratio is the impedance of a series resistor and capacitor. If the reactance X_C of the capacitor is 300 Ω and the resistor $R = 400$ Ω, the series impedance is 500 Ω.

The angles between the various pointers in Figure 1.16 are called *phase angles*. Sine waves that peak at different times are shifted in phase. To discuss phase relationships, the voltages must be sine waves at the same frequency. The current in a capacitor is always shifted 90° from the voltage across the capacitor. The current is said to lead the voltage. The voltage across a resistor is always in phase with the current. There is no phase shift in a resistor.

The RC Low-Pass Filter

A low-pass filter is a circuit that attenuates high-frequency sine waves and passes low-frequency sine waves. This filter might be used to limit high-frequency interference or reduce the high-frequency amplitude response of a voice amplifier. A first-order low-pass filter using an RC circuit is shown in Figure 1.17. This is exactly the same circuit we used to discuss the RC time constant in the previous section. This time our analysis involves sine wave voltages and not a step function.

The current flowing in the RC circuit is $I = V_{IN}/Z_{RC}$. The output voltage is $V_O = IX_C = V_{IN}X_C/Z_{RC}$. The ratio of output voltage to input voltage is

$$V_O/V_{IN} = X_C/Z_{RC} \qquad (1.3)$$

A low-pass filter is a voltage divider that changes the attenuated voltage depending on frequency. The ratio of output voltage to input voltage is called *gain* or *attenuation factor*. In this RC filter the gain is always less than 1.

To understand Equation 1.2 in more detail, let's look at the extremes of frequency. As the frequency rises, X_C gets smaller and Z_{RC} reaches a limiting value of R. At high frequencies the gain (attenuation) approaches $1/2\pi fR$. This means that the amplitude response falls off

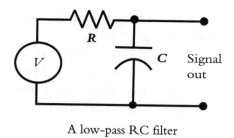

A low-pass RC filter

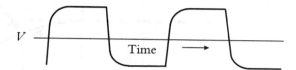

A square wave that has been filtered by an RC low-pass filter

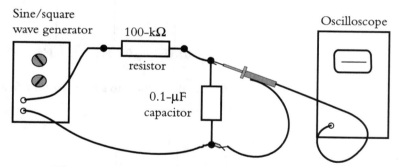

The wiring to the capacitor and the resistor

Figure 1.17 A first-order low-pass filter

inversely proportional to frequency. If f doubles, the amplitude is half its previous value.

At low frequencies X_C gets large compared to R. At the same time, Z_{RC} also gets greater. At low frequencies, Z_{RC} very nearly equals X_C and the gain ratio X_C/Z_{RC} approaches 1. This means there is little attenuation at lower frequencies. The frequency where $X_C = R$ holds a special significance. The gain at this frequency is 0.707. This is called the -3 dB (dB stands for decibel) frequency or cutoff frequency of the filter (see chapter 8).

The amplitude response of this filter as a function of frequency is given in Figure 1.18.

The amplitude response has been normalized so that the −3-dB point is at a frequency of 1 Hz. To find a resistor and capacitor value for a specific cutoff frequency f, first select a resistor value like 10 kΩ. If the cutoff frequency is 15 kHz (see chapter 8), then we can find C from the

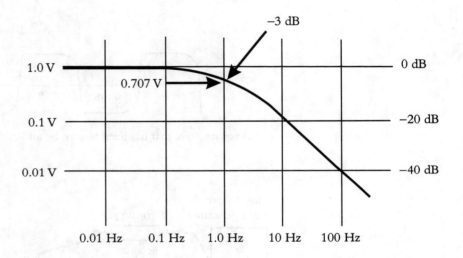

Amplitude response to sine waves for a first-order low-pass filter

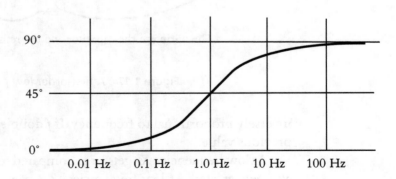

Phase shift of a first-order low-pass RC filter

Figure 1.18 The amplitude and phase response of a first-order low-pass RC filter

LEARNING CIRCUIT 7

Constructing and Observing a Low-Pass RC Filter

You will need (in addition to your measuring equipment):

1 0.001-µF capacitor

1 100-kΩ resistor

Construct a low-pass RC filter as in Figure 1.17. Let $C = 0.001$ µF and $R = 10$ kΩ. At 1 kHz set the amplitude level to equal 10 V peak. Measure the amplitude response across the capacitor at 1, 3, 5, 10, 15, 20, and 40 kHz. Use the oscilloscope as the voltmeter. Find the frequency where the amplitude is 7.07 V.

Change the function generator to a square wave at 2 kHz. Note the rounded character to the leading and falling edge of the square wave.

equation $X_C = R$. In this example $1/6.28fC = R = 10{,}000$. Solving for C, we obtain $C = 1/(6.28 \times 15{,}000 \times 10{,}000) = 0.0011$ µF. The response curve in Figure 1.17 is still applicable. Simply multiply the horizontal scale by 15 kHz.

The phase shift through this normalized RC filter starts off near 0 degrees at frequencies well below 1 Hz. This is because the reactance of the capacitor is much higher than the resistor and the capacitor performs like an open circuit. Well above the cutoff frequency, the capacitor acts like a short circuit and the current is limited by the resistor. This means that the current is nearly in phase with the source voltage. The voltage across the capacitor lags this current by 90°. At frequencies well above the cutoff frequency, the phase angle between the input and the output voltages approaches 90° lagging. At the cutoff frequency, the phase shift is 45°. The phase shift for this first-order filter is shown in Figure 1.18.

There is a close relationship between phase shift and attenuation slope. In this RC filter, the amplitude falls off proportional to frequency above the cutoff frequency. If two RC circuits were to contribute to the attenuation, the amplitude would fall off proportional to frequency squared. In this situation the phase shift would double. In general, phase shift is closely related (proportional) to the attenuation slope.

The RC High-Pass Filter

When the roles of the resistor and capacitor are reversed in Figure 1.17, the output voltage is sensed across the resistor. The frequency and phase response are mirror images of the low-pass filter. This filter attenuates low frequencies and passes high frequencies. The circuit and the amplitude and phase response are shown in Figure 1.19.

The square wave response of a first-order high-pass filter is shown in Figure 1.20.

The voltage across the resistor plus the voltage across the capacitor must equal the square wave voltage. In other words, the two top curves in Figure 1.20 add up to a square wave. The leading edge of the square wave comes through immediately in a high-pass filter. The output voltage falls as the capacitor charges up. After the leading edge is coupled, the voltage waveform follows an exponential curve. The voltage drops to 37% of initial value in 1 time constant equal to the product *RC*.

A high-pass filter can be used to block an average offset voltage. The filter allows changing voltages to pass. An application of this filter might be to reduce the bass or low-frequency response in an audio amplifier. It might be used to pass a high-frequency carrier signal and reject an audio signal.

 LEARNING CIRCUIT 8

Constructing and Observing a High-Pass Filter

Use the low-pass filter you constructed in Learning Circuit 7, but reverse the positions of the resistor and the capacitor. Find the amplitude response at the same frequencies. At what frequency does the amplitude reach 0.707 V? Observe the square wave response at 2 kHz.

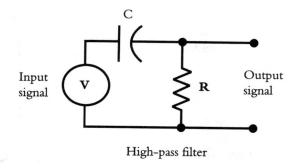

High-pass filter

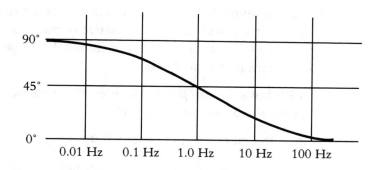

The normalized amplitude response to sine waves for a first-order high-pass filter

A normalized phase response for a first-order high-pass filter

Figure 1.19 The amplitude and phase response of a normalized first-order RC high-pass filter

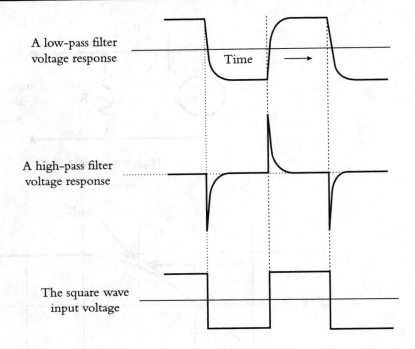

Figure 1.20 The response to a square wave for a low-pass and a high-pass filter

SELF-TEST

1. The terminals of the six individual cells of a 12-V battery are exposed. All the cells are in series. If the negative terminal is labeled 0 V, what are the voltages at the other cell terminals? What happens if the positive terminal is labeled 0 V?

2. Two 9-V batteries are placed in series. If the connecting point is 0 V, what are the other two voltages?

3. What is the maximum potential difference in problem 2?

4. Three resistors are in series. Their resistances are 910 Ω, 2.2 kΩ, and 3.3 kΩ. What is the total resistance?

5. Two resistors are in parallel. Their resistances are 1 kΩ and 10 kΩ. What is the total resistance?

6. Two resistors 510 kΩ and 1.2 MΩ are in parallel. What is the parallel resistance in units of kΩs and MΩs? (*Hint:* First rewrite the resistor values using the same units.)

7. A 10-kΩ resistor measures 5% high or 10.5 kΩ. Show that a parallel resistor of 220 kΩ will reduce the resistor value to near 10 kΩ. What is the remaining error expressed as a percentage of 10 kΩ?

8. 20 V dc is placed across a 10-Ω resistor. What is the current? What is the power dissipated? What is the current direction?

9. In problem 8, what happens if the battery is reversed in direction?

10. 100 V ac is placed across a 2-kΩ resistor. What is the current? What is the power dissipation?

11. 10 V is placed across a resistor of 0.1 Ω. What is the current? What is the power dissipation?

12. 500 Ω and 1,000 Ω are in series across a 15-V battery. What is the current in the resistors?

13. In problem 12 the 500-Ω resistor connects to the negative terminal of the battery. If the negative terminal of the battery is at 0 V, what is the voltage at the junction between the two resistors?

14. In problem 13, if the positive terminal of the battery is at 0 V, what is the voltage at the junction between the two resistors?

15. Solve problem 13 if the resistor values are doubled.

16. A 12-V battery is loaded with a 1.2-Ω load resistor. The voltage drops to 11.92 V. What is the source resistance?

17. A 15-V dc power supply has a voltage divider consisting of three 2-kΩ resistors. What are the voltages at the two junctions? What is the source resistance at these two points?

18. Use the circuit of Figure 1.9. If R_1 is 300 Ω and R_2 is 100 Ω and the voltage is 10 V, what is the voltage at the junction of the two resistors?

19. The voltage in Figure 1.9 is 20 V. If the attenuated voltage is 6 V, what are the two resistor values if the current drawn is 1 mA? What are the resistor values if the current drawn is 5 mA?

20. A 9-V battery drops to 8.8 V when a 100-mA load is applied. What is the internal resistance?

21. A 12-V battery supplies three lamps in parallel. The lamps have resistances of 2 Ω, 3 Ω, and 5 Ω. What is the total current? *Note:* The filaments of a lamp are made from tungsten. The cold resistance of the filament is much lower than the hot resistance. This problem assumes that the given resistance values occur when the lamps are illuminated.

22. What is the total capacitance when 0.01 µF is paralleled with 0.1 µF?

23. What is the total capacitance when 330 pF is paralleled with 0.002 µF?

24. What is the total capacitance when 0.001 µF is in series with 500 pF?

25. A capacitor of 2 µF has a voltage of 15 V. What is the charge stored?

26. In problem 25 a current of 2 mA flows for 1 ms. What is the charge?

27. A capacitor of 0.001 µF is charged for 10 µs with a current of 20 mA. What is the voltage on the capacitor?

28. What is the time constant when the resistor is 100 kΩ and the capacitor is 0.01 µF? How long are 2 time constants?

29. What resistor forms a 0.1-second time constant with a 0.05 µF capacitor?

30. An RC circuit reaches 95% of final value in 0.1 second. What is the RC time constant?

31. An RC low-pass filter has a cutoff frequency of 20 kHz. If $R = 10$ kΩ, what is C?

32. In problem 31, what is the attenuation of a 1-MHz signal?

33. A designer wants to use a 0.01-µF capacitor for an RC low-pass filter at 20 kHz. What resistor value must he use?

34. A high-pass filter has a cutoff frequency of 10 kHz. The R is 100 kΩ. What is the capacitor value?

35. A high-pass filter is formed using 0.01 µF and 150 kΩ. A step voltage of 10 V is applied. How much time elapses before the voltage drops to 3.7 V?

36. In problem 35, how much time is required before the voltage drops to 0.5 V?

37. A high-pass filter has a cutoff frequency of 10 Hz. If the resistor is doubled in value, what change in capacitor value must be made to maintain the same cutoff frequency?

ANSWERS

1. The voltages are 2, 4, 6, 8, 10, and 12 V. If the positive terminal is 0, the voltages are −2, −4, −6, −8, −10, and −12 V.

2. The voltages are +9 V and −9 V.

3. 18 V.

4. 7,410 Ω.

5. 909 Ω.

6. 353 kΩ or 0.353 MΩ.

7. The resistor is corrected to 9.976 kΩ. The error is 0.024 kΩ or 0.24%.

8. The current is 2 A. The power is 40 W. The current flows from plus to minus.

9. The current flows in the opposite direction.

10. The current is 50 mA. The power is 5 W.

11. 100 A. 1 kW.

12. 10 mA.

13. 5 V.

14. −10 V.

15. −10 V.

16. The voltage changed 0.08 V. The current changed 9.93 A. The source impedance is 0.008 Ω.

17. The voltages are 5 V and 10 V. The source impedance in both cases is 2 kΩ in parallel with 4 kΩ. This is 1.33 kΩ.

18. 2.5 V.

19. 6 kΩ and 14 kΩ. 1.2 kΩ and 2.8 kΩ.

20. 2 Ω.

21. 1.2 A.

22. 0.11 µF.

23. 0.00233 µF.

24. 333 pF.

25. $Q = CV = 2 \times 10^{-6} \times 15 = 30$ microcoulombs.

26. $Q = I \times t = 0.002 \times .001 = 2$ µC.

27. $Q = I \times t = 0.02 \times 10^{-5} = 0.2$ µC. $V = Q/C = 0.2$ µC$/0.001$ µF $= 200$ V.

28. $10^{-8} \times 10^5 = 10^{-3}$ sec. Two time constants = 2 ms.

29. $RC = 0.1$. $R = 0.1/5 \times 10^{-8} = 2$ MΩ.

30. Three time constants = 0.1 second. One time constant equals 0.03 seconds.

31. 10 kΩ $= 1/2\pi f C = 1/2\pi \times 20{,}000 \times C$. $C = 796$ pF.

32. The ratio of frequencies is 50:1. The attenuation factor is approximately 50.

33. $R = 1/2\pi f C$. $R = 1/6.28 \times 20{,}000 \times 10^{-8}$. $R = 796$ Ω.

34. 100 kΩ $= 1/6.28 \times 10{,}000 \times C$. $C = 159$ pF.

35. $RC = 10^{-8} \times 150{,}000 = 1.5$ ms.

36. Three time constants or 4.5 ms.

37. The reactance must also double. This means the capacitance is half the value.

2 Inductors, Transformers, and Resonance

Objectives

In this chapter you will learn:

- the way two more electrical components, the inductor and the transformer, work
- how an inductor is used in a filter
- what the L/R time constant is
- about the concept of resonance and how it is used
- about diodes and power supplies

The next two electronic components we will be studying are the inductor and the transformer. Building the Learning Circuits in this chapter will give you a chance to observe the way these components act in various combinations. Inductors and transformers have in common the fact that both of them use the magnetic field. If a review of the magnetic field would be helpful, now would be a good time to read "The Inductor and the Magnetic Field" in chapter 8.

Inductors are a fundamental component used in many electronics applications. At low frequencies, practical inductors are bulky and expensive, so they are not much used in low-frequency applications, such as audio. Their applications are in high-frequency power supply circuits, in power line filters, and in tuning and filtering in communications circuits. When combined with capacitors, inductors form resonant circuits. These circuits are important in the study of band selection in communications systems.

The transformer is a component that makes it possible to change ac power voltages to acceptable levels to operate circuits. Transformers also provide isolation so that many circuits can operate connected to different grounds. Transformers also allow circuits to be isolated from each other.

Figure 2.1 shows a coil of wire wrapped on a cylinder, in a geometry known as a *solenoid*. Inductors are usually, though not always, made in the shape of a solenoid. If a second coil of wire is wrapped around the solenoid, you have a transformer.

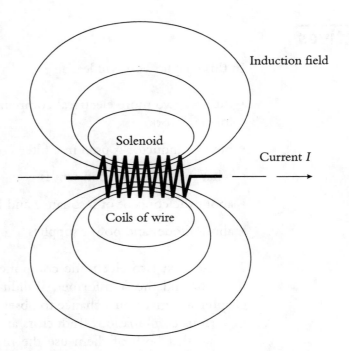

When current flows, an induction field is generated.

Figure 2.1 A solenoid—a coil of wire wrapped around a cylinder

When current flows in this coil, a magnetic induction field is created. The induction field is represented by a number of field lines that form closed curves around the path of current flow. These field lines thread through the center of the solenoid and complete their path on the outside of the coils. The number of magnetic lines represents the amount of induction flux that is developed. The number of lines crossing a given area is proportional to the magnetic intensity. The total number of lines is proportional to the current level and the number of turns. For a given current, the amount of induction flux can be increased by placing some magnetic material in the flux path.

High-quality inductors are often built using a bobbin—a mechanical structure for holding coils of wire—for the coils of wire and a ferrite core material that surrounds the bobbin. This construction forces the induction flux to cross a small air gap in the center of the core. The energy stored in the inductor is stored in this gap. Another construction uses a powered iron toroid (a donut shape) for the core. Coils of wire are simply threaded through the hole in the toroid. In this case the air gap is distributed throughout the magnetic material.

Induced Voltages

Consider the magnetic field in Figure 2.1. When a test coil of wire is moved in this magnetic field, a voltage appears on the ends of the coil. If the direction of motion is reversed, the voltage reverses its polarity. If the test coil is moved faster, the voltage increases. If the test coil is stationary and the current in the coil increases, there is a voltage on the test coil. If the current decreases, the voltage polarity reverses. The voltage is proportional to the rate at which lines of flux thread the test coil and to the number of turns. If the lines increase at 1,000 lines per millisecond, the voltage might be 10 V. If the lines decrease at 1,000 lines per millisecond, the voltage would then be −10 V. These are induced voltages.

Lines of flux thread the very coil that carries the current. When the current changes in the coil of Figure 2.1, there is a voltage at the terminals. This voltage is in the direction to oppose the change in current. The voltage is proportional to how fast the current is increasing. This fact is known as Lenz's law. If there is a changing current, there must be an induced voltage. This is known as Faraday's law.

Inductors and Inductance

The proportionality between the voltage and a changing current is known as *inductance*. In equation form, this relationship is written as

$$V = L \times \text{(the rate of change of current)} = L \times I/t \qquad (2.1)$$

where L is inductance in units of henries, I is current in amperes, and t is time in seconds.

An inductor is the component associated with this changing current. The symbol for an inductor is

Figure 2.2 shows this rise in current for a fixed voltage across an inductor.

If the voltage is 1 V and the inductance is 1 henry, abbreviated H, the current will rise at 1 A per second. Conversely, if the current rises at 1 A per second, the voltage is 1 V. It is interesting to compare this with a 1-farad capacitor. The voltage will rise at 1 V per second for a current of 1 A. Conversely, if the voltage rises at 1 V per second, the current is 1 A.

A 1-H inductor is not used very often in electronics. Typical inductor values range from a few microhenries to perhaps 100 millihenries. The millihenry is 0.001 H, abbreviated mH, and the microhenry is 0.000001 H, abbreviated μH.

The Reactance of an Inductor

When a sinusoidal current flows in an inductor, a sinusoidal voltage will appear across its terminals. The equation for a sine wave of current is

$$i = I \sin 2\pi ft \qquad (2.2)$$

The maximum rate of change of current is $2\pi fI$ where f is frequency in Hz and I is the peak current in amperes. The peak voltage is $2\pi fLI$

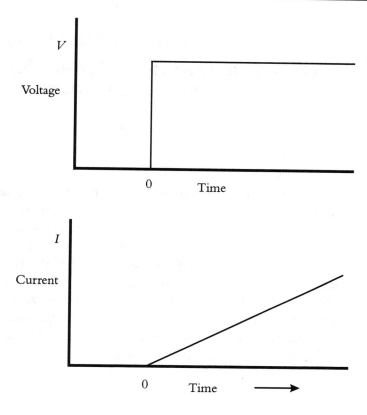

At the moment the voltage is applied across the terminals of an inductor the current starts to rise linearly. In an ideal inductor the voltage must be reversed in polarity to return the current to 0.

Figure 2.2 The current rise in an inductor

where L is the inductance in henries. The ratio of peak voltage to peak current is

$$X_L = 2\pi f L \qquad (2.3)$$

where X_L is called the *inductive reactance*. The unit of reactance is the ohm. The maximum rate of change of current for a sine wave current

LEARNING CIRCUIT 9

Observing the Rise in Current for a Fixed Voltage in an Inductor

You will need (in addition to your measuring equipment):

1 10-mH inductor

1 10-Ω resistor

To put together the circuit shown in Figure 2.3, use the construction layout at the bottom of the figure. When you use a 10-mH inductor, a voltage of 1.0 V dc across the inductor will cause a current to rise at 100 A/sec. If we limit the maximum current to 0.01 A, we cannot leave the voltage connected for more than 100 μs. We can do this with a square wave voltage that stays positive for 100 μs and reverses polarity for another 100 μs. This is a square wave at 5 kHz. A low-valued resistor in series with the inductor can be used to measure the current. The peak voltage across a 10-Ω resistor will be 100 mV. Use the oscilloscope to observe that this voltage is a triangle wave.

occurs when the current is 0. In an inductor, when the sinusoidal current is 0, the voltage is maximum. The rotating pointer system in Figure 2.4 shows the timing relationship between current and voltage.

In an inductor the voltage leads the current by 90° at all frequencies. Compare this with a capacitor, where the voltage lags the current by 90° at all frequencies.

Inductors in Series and Parallel

Inductors in series add. All of the units must agree. For example, to add 100 μH to 2 mH requires the 100 μH be converted to 0.1 mH. The sum is 2.1 mH. Inductors in parallel are treated the same as parallel resistors. The reciprocal of reactance is susceptance. The susceptances are first added together. The reciprocal of the total susceptance is the parallel inductance. Consider 3 mH in parallel with 2 mH. The reciprocals

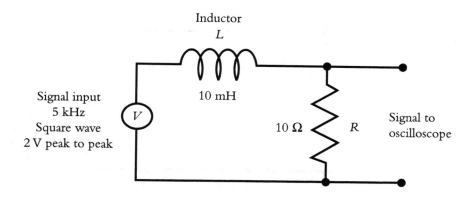

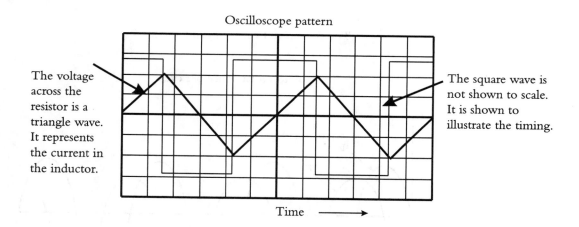

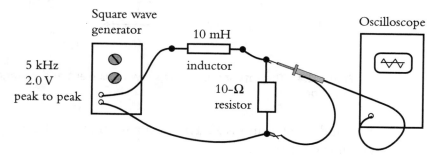

Figure 2.3 The rise and fall of current in an inductor

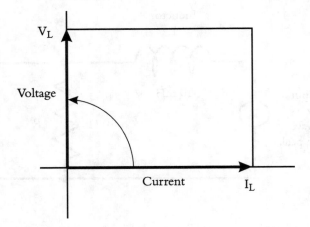

The current and voltage in an inductor.

The pointers rotate once per cycle.

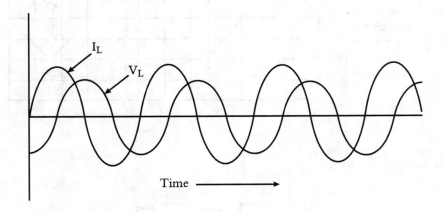

Figure 2.4 The rotating pointers showing current and voltages in an inductor

are ⅓ and ½. The sum is ⅚. The final answer is ⅚ = 1.2 mH. Again, all the units must agree.

The Resistor-Inductor (L/R) Time Constant

Figure 2.5 shows a step voltage applied to a series resistor–inductor (RL) circuit.

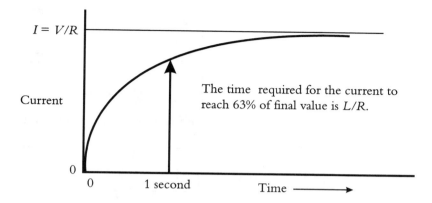

The rise in current follows an exponential curve.
In this figure the value of L/R is 1 second.

Figure 2.5 The rise in current for step function applied to series RL circuit

At the moment the voltage is applied, the current in the circuit is 0. The entire voltage appears across the inductor. Equation 2.1 requires that the current start to increase. This current results in a voltage drop across the resistor. This reduces the voltage across the inductor, which in turn reduces the rate at which current is rising. The current continues to increase until it reaches a limiting value determined by the voltage and the resistor.

This rise in current follows an exponential curve. This is the same as the voltage curve shown in Figure 1.14 for the RC time constant. The current reaches 63% of final value in a time given by the ratio L/R. To show that L/R has units of time, we can use Equation 2.1, which states that volts = (inductance × amperes)/time. This means inductance = (volts/amperes) × time. Ohm's law says that resistance = volts/amperes. Dividing inductance by resistance, the units volts/amperes cancel, leaving the unit of time. For L/R to equal time in seconds, the inductance must be in henries and the resistance in ohms.

The voltage across the resistor R in Figure 2.5 is a direct measure of the current. If a square wave voltage is used instead of the step function, the voltage waveform across the resistor is a filtered version of the square wave. This waveform is shown in Figure 2.6.

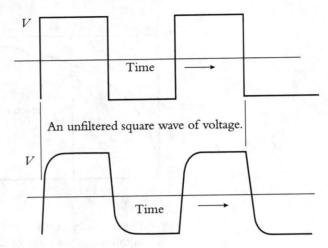

Figure 2.6 A square wave of voltage filtered by an RL circuit

The Impedance of a Resistor and Inductor in Series

Impedance is a sinusoidal concept. Our objective is to find the ratio of sine wave voltage to sine wave current in a series RL circuit. If the current is I, the voltage across the resistor peaks 90 electrical degrees after the voltage across the inductor. The rotating pointer system in Figure 2.7 shows these two voltages as sides of a rectangle.

The applied voltage V is represented by the diagonal of this rectangle. The peak voltage across the resistor is IR and the peak voltage across the inductor is IX_L. The peak voltage V is

$$V = \sqrt{(I^2 R^2 + I^2 X_L^2)} \tag{2.4}$$

Dividing both sides by I yields the impedance

$$Z = V/I = \sqrt{(R^2 + X_L^2)} \tag{2.5}$$

where Z is in ohms. The impedance Z depends on frequency as X_L increases linearly with frequency.

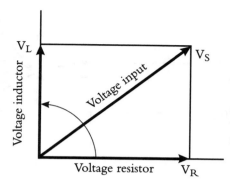

At time = 0 the voltage across the resistor is zero and the voltage across the inductor is at its maximum negative value.

The pointers rotate once per cycle.

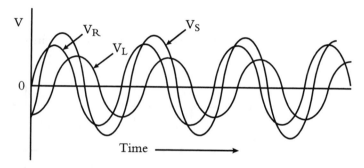

The three sine wave voltages displayed in time.

Figure 2.7 The rotating pointers showing voltages in an RL circuit

The RL Low-Pass Filter

The circuit in Figure 2.3 is a low-pass filter. When a sine wave of voltage V_{IN} is applied to the circuit, the voltage across the resistor is the output of the filter V_{OUT}. The ratio V_{OUT}/V_{IN} is the gain of the filter. In this filter the gain is always less than 1.

The current in the circuit is $I = V_{IN}/Z$. The voltage out is IR or $V_O = V_{IN}R/Z$. At dc the value of Z is R, so the gain is 1. When $R = X_L$, the impedance Z is equal to $R\sqrt{2}$. The gain at this frequency is equal to 0.707. This frequency is called the *cutoff frequency*. Above this frequency

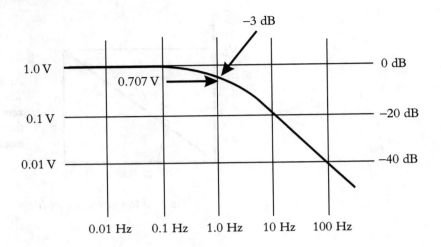

Amplitude response to sine waves for a first-order low-pass filter.

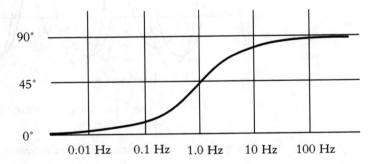

Phase shift for a first-order low-pass filter.

Figure 2.8 The amplitude and phase response of a first-order low-pass RL filter

the impedance is dominated by the reactance of the inductor, which rises linearly with frequency. This means that the gain falls off linearly with frequency. A plot of gain and phase shift for a 1-Hz low-pass filter is shown in Figure 2.8. This is an exact duplicate of the RC low-pass filter discussed in chapter 1. In the RC filter the output voltage is sensed at the capacitor, and in the RL filter the output voltage is sensed at the resistor.

The RL High-Pass Filter

When the resistor and the inductor are exchanged in Figure 2.5, the circuit becomes a high-pass filter. At high frequencies the inductor has a high reactance and does not affect the signal. At low frequencies, where the reactance is low, the signal is attenuated. The curves of gain and phase are the same as the RC high-pass circuit shown in Figure 1.19. The square wave response is the same as in Figure 1.20.

Small inductors are often placed in circuit leads to restrict the flow of current at high frequencies. This type of inductor can be as simple as threading a conductor through a small magnetic core (ferrite bead). The filtering action depends on the presence of a shunting impedance. If this impedance is not present, the filter might not function.

Later when we discuss resonant frequency you will see that inductors are not perfect. Above a certain frequency an inductor functions like a capacitor and the expected filtering action does not take place. All components have their limitations, but inductors have several weaknesses. Experience tells a designer when and how he can use an inductor. Inductors in the microhenry range work well above 1 MHz when the impedances are below a few hundred ohms.

The Series Resonant Circuit

Resonance occurs often in mechanical systems. Examples are the ringing of a big bell, the ringing of a wineglass, or the swinging of a pendulum. The three things that are needed to create a resonant response are a source of energy and ways to store potential energy (energy of position) and kinetic energy (energy of motion).

In an electric circuit the energy stored in a capacitor can be considered potential energy. This energy depends on the position of charge. The energy storage in an inductor depends on current flow. This energy can be considered kinetic. If the circuit can be arranged so that current can flow between the capacitor and the inductor, then the ringing associated with resonance will occur.

Figure 2.9 shows a series RLC circuit connected to a sinusoidal voltage. The impedance of this circuit can be determined by assuming a sinusoidal current flow and solving for the voltage. The ratio of voltage to current is the impedance.

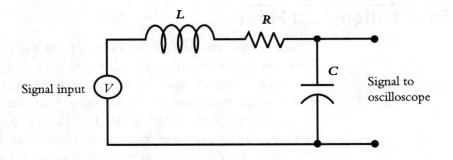

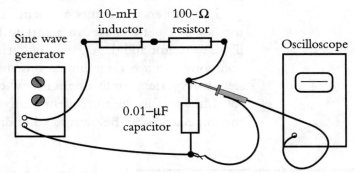

The wiring to the resistor, the capacitor, and the inductor.
This circuit is a second-order low-pass filter.

Figure 2.9 A series RLC circuit

The voltage across the inductor is IX_L, the voltage across the capacitor is IX_C, and the voltage across the resistor is IR. The voltage across the capacitor lags the voltage by 90° and the voltage across the inductor leads the current by 90°. The voltage across the resistor is in phase with the current. The pointer system in Figure 2.10 shows this timing relationship. The voltage across the resistor is the reference voltage, and it points to the right.

The voltage pointers for the capacitor and inductor point in opposite directions. This means that these voltages subtract. In Figure 2.10 the voltage across the inductor dominates and the result is a net inductive reactance. We can now construct a pointer that is the sum

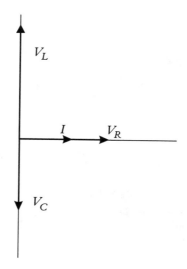

Figure 2.10 The pointer system applied to a series RLC circuit

of these three voltages. The difference voltage $IX_L - IX_C$ is one side of a rectangle. The other side is IR. The length of the diagonal is $V = \sqrt{I^2R^2 + I^2(XL - XC)^2}$. The ratio V/I is the impedance of the circuit or

$$Z = V/I = \sqrt{R^2 + (X_L - X_C)^2} \qquad (2.6)$$

At the frequency where $X_L - X_C = 0$, the impedance becomes R. This is known as the *resonant frequency*. At frequencies below resonance, the capacitive reactance dominates, and at frequencies above resonance the inductive reactance dominates. If the resistance were 0, the series resonant circuit would be a short circuit at the resonant frequency. In practice this resistance can only approach 0. There are several factors, including the dc resistance of the coil, that add to the effective resistance. A near short circuit at one frequency allows this circuit to be used as a selective filter. The impedance of the RLC series circuit as a function of frequency is shown in Figure 2.11.

The resonant frequency can be determined by letting $X_L = X_C$ or $2\pi fL = 1/2\pi fC$. Solving for f yields

$$f = 1/(2\pi\sqrt{LC}) \qquad (2.7)$$

where f is in hertz, L is in henries, and C is in farads.

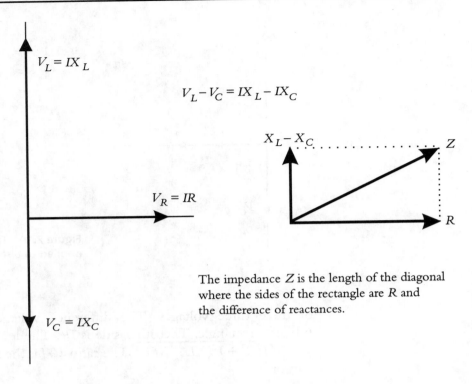

In this example the voltage across the capacitor subtracts directly from the voltage across the inductor. At resonance the voltage across the capacitor equals the voltage across the inductor.

Figure 2.11 The impedance of a series RLC circuit

More About the Series Resonant Circuit

If we use the RLC circuit as a low-pass filter, it offers an opportunity to demonstrate a few advanced topics in electronics. These topics involve circuit stability, overshoot, and ringing. See Figure 2.12. Many circuits exhibit the character of an RLC circuit, yet they are designed from different components. The circuits are different, but the response can be the same as an RLC response.

Above the resonant frequency the current in this RLC circuit is delayed by the inductor by 90°. The voltage across the capacitor is delayed by 90° from the current. The result is that the output voltage well above the resonant frequency is shifted in phase by nearly 180°.

◦⌇◦ LEARNING CIRCUIT 10 ◦⌇◦
Observing the Response of an RLC Circuit

You will need (in addition to your measuring equipment):

1 10-mH inductor

1 100-Ω resistor

1 0.01-μF capacitor

Turn back and look again at Figure 2.9. The RLC filter in this figure uses a 10-mH inductor, an 0.01-μF capacitor, and a 100-Ω resistor. This circuit has a gain of 1 at dc and low frequencies. As the frequency of the input sine wave nears the resonant frequency, the current rises as the impedance drops. At resonance the current is maximum, and the voltage across the capacitor can be quite large. The voltage is limited by the value of the resistor. Assume the input voltage is a 1-V sine wave. If the reactance of the capacitor at resonance is 1,000 Ω and the resistor is 100 Ω, the output voltage will be 10 V. The output voltage amplitude is maximum near the resonant frequency.

Connect a sine wave generator to these three components and you can observe this amplitude response across the capacitor. If you were testing a circuit and you saw this response, it could be a sign of instability, which is not desirable. In our circuit, if the resistor is increased, the peaking effect is reduced. In a circuit where the equivalent resistance goes to 0, the result will be an oscillator.

When the circuit in Figure 2.9 is excited by a square wave at 1 kHz, the result is ringing. This ringing voltage is shown in Figure 2.12. If the resistor is increased, the ringing will diminish. There is a point where the square wave response has 2 or 3 percent of overshoot. This is an optimum response. In designing circuits where stability is a consideration, this is how the circuit should respond. This demonstration illustrates the power of a square wave. By observing the square wave voltage response, we can infer the response of the circuit to a wide range of sine waves. Change the resistor to 1,000 Ω and observe that the ringing is reduced. At $R = 1,400$ Ω, the response is optimum.

The amount of ringing to a square wave is related to a factor called *damping*. A damping factor of 1 implies that there is no overshoot. A

(Continued)

damping factor of 0.7 implies an overshoot of a few percent. A damping factor of 0.3 implies a significant overshoot and many ring-down cycles. See Figure 2.12. In theory the ringing lasts forever. In practice a number of cycles can be seen, depending on the damping factor. A damping factor of 2 implies a very sluggish response.

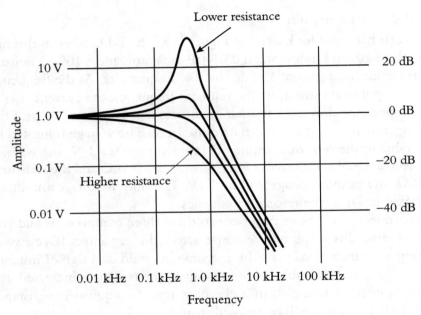

Amplitude response of a second-order filter for various resistance values

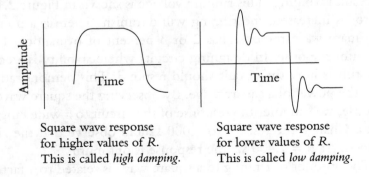

Square wave response for higher values of R. This is called *high damping*.

Square wave response for lower values of R. This is called *low damping*.

Figure 2.12 The amplitude response and square wave response of a series RLC circuit

The attenuation at frequencies above the resonant frequency increases proportional to the square of frequency. Another way of saying this is that the amplitude falls off at 40 dB per decade. If the output is at 1 V at 50 kHz, the response is 0.01 V at 500 kHz. You can check this using the circuit components in Figure 2.9.

The Parallel Resonant Circuit

A parallel resonant circuit is shown in Figure 2.13. When a voltage is impressed on this circuit, reactive current flows in both the capacitor

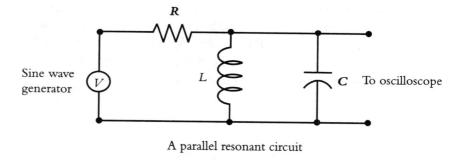

A parallel resonant circuit

The resonant frequency is $\dfrac{1}{2\pi \sqrt{LC}}$

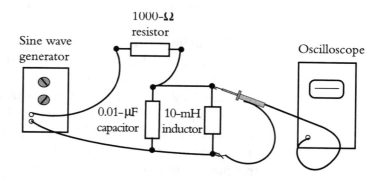

Figure 2.13 A parallel resonant circuit

LEARNING CIRCUIT 11

Observing the Response of a Parallel Resonant Circuit

The same components used in Learning Circuit 10 can be used to form a parallel resonant circuit as shown in Figure 2.13. Place this parallel LC circuit in series with a 1,000-Ω resistor across a sine wave generator. Rearrange the circuit so the resistor goes to the signal reference or signal generator ground. Set the voltage to 10 V peak. Note the voltage across the resistor with the oscilloscope. At the resonant frequency the voltage will drop to a low value. Note the attenuation factor. Add a 10-Ω resistor in series with the inductance. Note that the attenuation factor is reduced. Plot the peak output voltage as a function of frequency with and without the added 10-Ω resistor. This change in attenuation is explained in the next discussion.

and the inductor. These currents are of opposite sign and they subtract. At the resonant frequency the current in the capacitor equals the current in the inductor. The current supplied by the external voltage is 0. The current for the inductor is supplied by the capacitor, and the current for the capacitor is supplied by the inductor. Current flows, but it is not supplied by the external voltage. An ideal LC parallel circuit is an open circuit at its resonant frequency. This circuit can be used to limit current flow at one frequency. This parallel LC circuit is often called a *frequency trap.*

The inductor in the parallel resonant circuit has some resistance. To see the effect this resistance has on the circuit impedance, it is convenient to consider the energy lost in the resistance. As an example, consider 10 V across a parallel resonant circuit where $L = 10$ mH and $C = 0.10$ µF. The resonant frequency is 15.92 kHz. The reactance of the inductor is 1,000 Ω. The current that circulates in the resonant circuit for $V = 10$ V is 10 mA. If the resistance in the inductor is 10 Ω, the power dissipated is $I^2R = 1$ mW. This power must be supplied by the 10-V source. The power is $VI = 1$ mW. But $V = 10$ V and the current I must equal 0.1 mA. The impedance given by Ohm's law is

$V/I = 10/0.0001 = 100$ kΩ. This is a high impedance compared to the reactance. If the series resistor were only 1 Ω, the impedance would increase to 1 MΩ.

Transformers

A transformer is a circuit component consisting of coils of wire encircling a path for magnetic flux. Refer back to the coil shown in Figure 2.1. If a second coil is added to this inductor, it becomes a transformer.

At frequencies above 1 MHz, the solenoid does not need a core; the magnetic path can be in air. In applications at power frequencies (60 Hz), however, we must provide a magnetic material as a path for the magnetic flux.

When a voltage is applied to a coil of wire, the result is an increasing induction field. The induction flux increases as long as the voltage is present. This induction flux or B field threads through the coil. This induction field exists independent of whether there is magnetic material present or just air. The induction flux requires an associated magnetizing current that also increases with time. At 1 MHz the voltage is positive for a half-cycle, or ½ µs. The magnetizing current must increase for ½ µs before the voltage reverses polarity and the current starts to reverse direction. At 60 Hz the current would have to increase for 8 ms, which is 16,000 times longer. If the maximum current at 1 MHz were 10 mA, the maximum current at 60 Hz would be 160 A. This level of current is unacceptable for a 60-Hz transformer. This is the reason why a magnetic core must be provided.

When a voltage is applied to a coil of wire, the current that flows is called the *magnetizing current*. This reactive current is associated with an inductance called the *magnetizing inductance*. The presence of iron in the magnetic path reduces the amount of magnetizing current that is required. If the induction flux follows a magnetic material along its entire path, the magnetizing current can be significantly reduced. The reduction factor is known as the *permeability of iron*. In the previous example, a permeability of 10,000 would reduce the current from 160 A to 16 mA, an acceptable level.

The induction flux created by a voltage across a coil of wire follows the magnetic path as this path stores the least amount of magnetic field energy. If a second coil is wrapped around the first coil, then any changing induction flux threads both coils. The voltage on this added coil depends on how rapidly the induction flux is changing and on the number of turns. If both coils have 1,000 turns, then the voltage on both coils will be identical. If the second coil has 500 turns, the voltage will be half. This transfer of voltages between coils of wire is called *transformer action*. The coil receiving the initial voltage is called the *primary coil*. All other coils are called *secondary coils*.

There is a maximum induction flux that can be supported by a magnetic material. When a voltage is applied to a coil on a transformer, the induction flux starts to rise. If the voltage persists for too long, the core saturates. This means the permeability drops. In this situation the magnetizing current must increase to support the changing induction field. Transformers are functional as long as the limits of the induction field are not exceeded.

The symbol for a transformer is shown in Figure 2.14. The bars between the coils represent the iron in the core. This symbol is sometimes misleading, because it does not represent the actual construction of the transformer. For example, there is capacitance between the primary coil and the secondary coil. The capacitance is not symmetrically distributed, as the symbol might suggest, but is largely between the last turns of the primary coil and the first turns of the secondary coil.

Power transformers (60 Hz) are built by winding the coils on a bobbin. The core material is then added by interleaving iron laminations. These laminations provide a magnetic path for the magnetic flux. Lam-

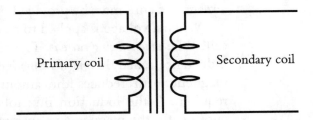

The ratio of primary to secondary voltage is the same as the turns ratio.

The vertical lines imply that the transformer has a magnetic core.

Figure 2.14 The transformer symbol

inations are required because the magnetic field at 60 Hz can only penetrate a short distance into iron. A typical lamination thickness at 60 Hz is 15 mils or 0.015 inches. 1 mil is a thousandth of an inch. At 400 Hz the lamination thickness is about 5 mils.

There are many applications where transformers must operate at higher frequencies. The core material for these transformers is made from a material known as *ferrite*. Ferrite is a powdered magnetic material mixed with a filler so that the mixture is an insulator. The mixture is then sintered in an oven to form a very hard material. (Sintering is a heating process that melts the mixture.) The bits of magnetic material are small enough so that magnetic field can penetrate into the magnetic material at high frequencies. In other words, ferrite materials have permeability at high frequencies.

The intensity of the magnetic induction field has units of teslas. In transformer design the unit used is the gauss. One tesla is equal to 10,000 gauss. The number of lines of magnetic induction flux is simply the magnetic intensity (gauss) times the cross-sectional area in cm^2. Magnetic materials can support a limited number of field lines. If the cross-sectional area is increased, the number of field lines that can be supported is also increased. Air can support an unlimited number of field lines. Iron begins to saturate above 16,000 lines per cm^2. Saturate here means the material loses its permeability and begins to behave magnetically like air. Ferrites often saturate above 8,000 lines per cm^2. When a magnetic material is saturated, it behaves like air.

When the number of field lines that thread through a one turn of wire increases at a linear rate, a steady voltage will appear at the ends of the turn. When a coil of wire replaces the single loop, the voltage is proportional to the number of turns in the coil. When a voltage is placed on a single turn of wire, the number of field lines that thread the coil must change at a fixed rate. If the applied voltage is fixed and if the number of turns is increased, the voltage per turn is decreased. This reduces the rate at which the field lines must change.

As an example of how a transformer works, consider a voltage applied to a primary coil of 500 turns. The core cross-section is 5 cm^2. Assume the induction flux increases at 12,000 lines per ms. After 4 ms the number of lines is 48,000 lines. This is 9,600 lines per cm^2, well within the saturation limits of the magnetic material. At this point the voltage is changed to −25 V. The field lines begin to decrease. In 8 ms the number of lines is again 0 and in 12 ms the field lines have reversed direction and the number is −48,000 lines. At this point the voltage is again reversed and in 16 ms the number of field lines is again 0. One cycle has taken 16 ms, which is a frequency of 60 Hz.

If the number of turns in the coil were decreased to 250, the number of field lines would have to change at twice the rate. This means the maximum number of lines per cm^2 would be 19,200. This is enough to saturate the magnetic material. To avoid saturation the core area could be increased or the number of turns increased.

Transformer Voltages and Currents

The secondary voltages of a transformer are often lower than the primary voltage. The ratio of secondary turns to primary turns determines the voltage. If the primary has 1,000 turns and is 120 V, a secondary coil of 100 turns will be 12 V. If a load is placed on this secondary coil, current will flow as determined by Ohm's law. For example, if the load is 12 Ω, the current is 1 A. The power dissipated is 12 W. This means that the 120-V primary voltage must supply 0.1 A to the primary coil. Viewed from the primary side of the transformer, a voltage of 120 V and 0.1 A represents an impedance of 1,200 Ω. A 12-Ω load on the secondary represents a 1,200-Ω load to the primary voltage. The ratio of load impedance to input impedance is 100:1. This is the square of the turns ratio.

If a transformer has two secondary coils, then each coil can be independently loaded. The current waveforms demanded by the secondary coils are reflected on the primary by the turns ratio. For example, if the primary voltage is 120 V and the secondary voltage is 12 V, a 3-A pulse of current on the secondary is supplied by a 0.3-A pulse of current in the primary coil.

The resistances of the primary and secondary coils are adjusted by the designer so that power losses in the coils are about equally divided. Since the primary coil must handle all the power, this coil occupies about half the available volume. It is common practice to wind the primary coil next to the core. Coils that are intended to supply a higher current have the lowest resistance.

Many small power transformers have high coil resistances. The result is that magnetizing current flowing in the resistance of the primary coil modifies the voltage waveform appearing on the primary coil. This change in voltage waveform appears on the secondary coils.

The magnetizing current is apt to be greatest when the voltage waveform is at a zero crossing. In most dc power supplies the current is supplied in short pulses. This current flows in the coil resistances of the transformer and further reduces the voltage available to the load. In short, small power transformers (up to 10 W) have many practical limitations. The loaded waveforms are often very different than sine waves.

Diodes

A diode is a component that allows current to flow in one direction (see chapter 8). The symbol for a diode is

The current flows in the direction of the arrow. This is called the forward direction of the diode. In the reverse direction a diode will withstand a high voltage before conducting. There are many types of diodes available to accommodate high current, high reverse voltage, and high-frequency applications. The forward drop in a silicon diode is about 0.6 V. Diodes are often used in groups of four. This configuration is known as a *bridge rectifier*. A package of four diodes is available as a single component with four connections. We will discuss the diode in detail in the next chapter. For the moment, we need the diode to discuss power supplies.

DC Power Supplies

Most of the circuits in electronics function from dc power supplies, but batteries are the sole voltage source in some applications. Sometimes rechargeable batteries are used, and the batteries are recharged whenever utility power is available. In this section we will look at dc voltages derived from utility power, not batteries.

There are many adapters available commercially. An adapter is a transformer that reduces the utility voltage to a safe level. Some adapters provide a dc output, while others provide an ac voltage. An adapter is used so that the equipment does not plug directly into a utility receptacle. An example might be the charger for a cell phone or the power for a telephone set. An ac adapter is a convenient source of ac power for many of our experiments. The low voltage from the adapter can be used without fear of electrical shock. An ac adapter is used in Figure 2.15 to demonstrate how the diodes convert ac to dc.

The output voltage V follows the ac source voltage on the positive half of the power cycle. A 10,000-Ω resistor is used as a load so that the voltage is 0 when the diode is not conducting. This is called *half-wave rectification*. If the adapter is rated 10 V unloaded, the peak output voltage is 14.14 V less a diode drop of 0.6 V. This is approximately 13.54 V.

This half sine wave can be used to charge a capacitor. This added capacitor is shown in Figure 2.16.

The capacitor charges to the peak ac voltage, but the capacitor cannot discharge back into the transformer as the diode prevents current flow in that direction. Load resistor R_L draws current from the capacitor, causing the voltage to sag each cycle. The ac voltage rises in each

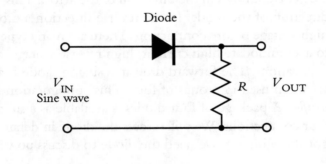

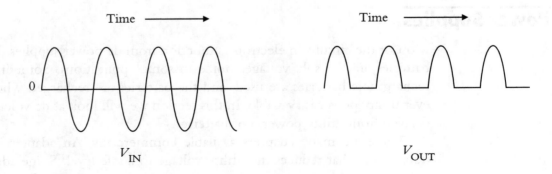

Figure 2.15 The action of a diode on a sine wave voltage

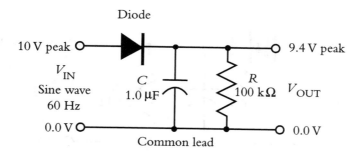

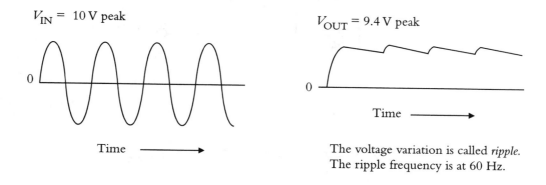

The average dc voltage output is about 8 V.

Figure 2.16 A diode and a capacitor providing a dc voltage

cycle to recharge the capacitor. The charging current flows when transformer voltage exceeds the capacitor voltage plus the diode voltage drop.

The capacitors used in this application are called *filter capacitors*. At 60 Hz these capacitors are usually polarized, which means a voltage can only be applied in one direction. The dielectric is an electrolyte that has a very high dielectric constant for one polarity of voltage. These capacitors are called *electrolytics*. If the polarity is reversed, the capacitor will conduct and probably overheat. The polarity markings for an electrolytic capacitor are usually on the case of the capacitor. Sometimes only one of the polarity signs is displayed.

The problem with the circuit in Figure 2.16 is that a dc current flows in the transformer secondary winding. This will tend to saturate

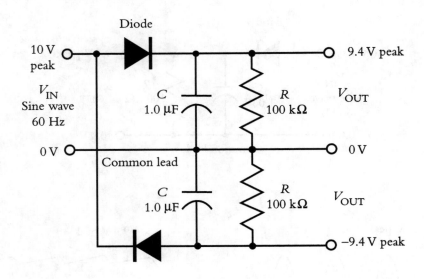

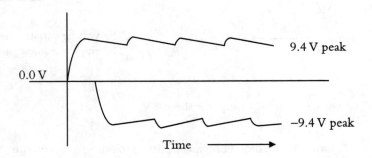

The two dc voltages developed by two diodes.
The ripple frequency is 60 Hz.

For a practical power supply, the capacitors must be at least 100 μF.
If the resistors are lower in value, the ripple voltage will increase.
A power supply does not normally need internal load resistors.

Figure 2.17 The use of two diodes to form two dc power supply voltages

the transformer core, and the transformer may draw excessive magnetizing current. This excessive current can overheat the transformer. To avoid this problem, a second diode can be used to supply a second power supply. If the transformer supplies equal current on the other half-cycle, the transformer will not saturate. A power supply with a positive and negative dc voltage finds wide acceptance in circuit design. This circuit is shown in Figure 2.17.

⌇ LEARNING CIRCUIT 12 ⌇
Building a DC Power Supply

You will need (in addition to your measuring equipment):

an 18-V ac adapter (do not use a dc adapter)

circuit board

soldering iron and solder

bus wire

press fit pins

2 2.2-kΩ and 4 1-kΩ resistors

2 100-μF, 35-V capacitors

2 1N1002 diodes

This Learning Circuit, as well as future Learning Circuit exercises, requires an ac adapter. Before proceeding further, open the secondary cable of the adapter with a sharp knife and solder the conductors to terminals on a circuit board. The ac voltage should be about 18 V rms. You can measure this ac voltage using a multimeter on the ac voltage range.

The construction of the power supply circuit in Figure 2.17 is shown in Figure 2.18. The two diodes and two electrolytic capacitors should be mounted on terminals that press fit into the board. The power diodes can be a 1N1002 or 1N1003; this is a readily available component. The 100-μF capacitors can be rated any voltage greater than 35 V. A larger capacitor value is acceptable: 200 μF.

Be careful to observe polarity on the capacitors and the forward direction on the diodes. The diode bar goes to the plus capacitor terminal.

(Continued)

The two dc voltages are the ac voltage times 1.414, less a diode drop of 0.6 V. This means the dc voltages should be about 25 V plus and minus.

Observe the dc voltages using the oscilloscope with the input shield connected to the 0 of potential or common. This is the conductor between the two capacitors. Place 1,000-Ω 1-W resistors between the two dc voltages and common and observe that the dc voltage drops slightly. Place a second 1,000-Ω 1-W resistor across each supply and again note the voltage change. Calculate the source impedance. Calculate the power dissipated in each resistor. Notice the ripple voltage (peak-to-peak ac voltage) across the capacitors for one set of resistors and for the second set. To observe power supply ripple voltage on your oscilloscope, it may have to be in the ac coupling mode.

When you build this power supply, do so on one end of the circuit board and leave room for other circuits. We will be using it in future Learning Circuit exercises. For ease in adding later circuits, use three long conductors that go across the board. The top conductor is +25 V, the middle conductor is 0 V (common or ground), and the bottom conductor is −25 V.

Schematics and Construction Diagrams

You may have observed that Figures 2.17 and 2.18 are very different drawings, yet they both describe the same circuit. Figure 2.17 is what is called a *schematic*. It shows only the actual components used in the circuit. Figure 2.18 is a construction diagram, and shows how these circuits could be actually constructed by being laid out on a circuit board. It is only one of many possible layouts or arrangements. The schematic is unique to the circuit; the construction layout is simply a suggestion of a possible way the circuit might be constructed. In several previous drawings (e.g., Figures 2.9 and 2.13), I drew them both on the same page, with the schematic above and the construction diagram below. Now that you are building larger and more complicated circuits, I will need to provide you with separate drawings—there will not be enough space to include them both in one figure.

Inductors, Transformers, and Resonance

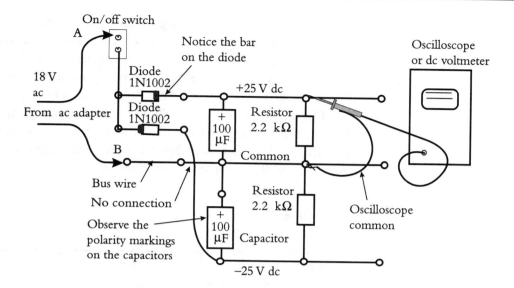

To construct this circuit, press pins into the circuit board in the arrangement shown. Solder the components to the pins. Use small pieces of bus wire to make all other required connections. Connect the adapter output leads to the circuit at A and B. Build the circuit on the left side of the board leaving room for more circuitry. Electrical connections are limited to the pins. Do not make any other connections.

The resistors are not needed for the power supply. They are load resistors that draw current from the capacitors so that you can observe power supply ripple. You may remove the resistors after you see the ripple.

Move the oscilloscope probe to the negative power supply voltage to observe the ripple and the dc value. Leave the oscilloscope probe common connected to the power supply common when you make this measurement.

This power supply circuit will be used in many of the following Learning Circuits. Leave it connected when you are through with this lesson.

Figure 2.18 The construction of a dual dc power supply

Transformer secondary coils are often wound with a connection to the midpoint. This midpoint on a secondary coil is called a *centertap*. If the total coil voltage is 20 V, the winding is often referred to as 10 V-0-10 V. The midpoint is often used as the zero voltage or reference conductor for the circuits that use this winding for power. If the centertap is at 0 V, the top of the coil is positive when the bottom of the coil is negative. The positive voltage can charge a capacitor through one diode while the negative voltage can charge a second capacitor through a second diode. On the next half-cycle the voltages change polarity, and two more diodes can be used to charge the same two capacitors. Figure 2.19 shows this four-diode arrangement where the two capacitors are charged once each half-cycle. This is called a full-wave rectifier circuit, as current flows in each half-cycle. This is the most efficient way to use the transformer. Unfortunately, most available adapters (transformers) are not centertapped. This will not cause us a problem in our Learning Circuits.

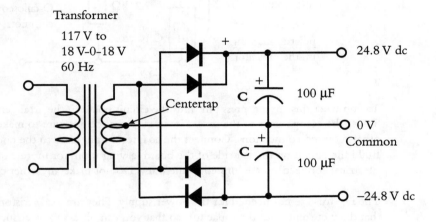

The ripple frequency is 120 Hz.

The four rectifiers are available as one component. The component is called a *bridge rectifier*. The terminals are marked ac, ac +, and –.

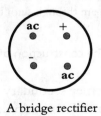

A bridge rectifier

Note: The adapter transformer does not have a centertap.

Figure 2.19 A full-wave centertap rectifier circuit

The circuit in Figure 2.19 can be used without a transformer centertap. The two capacitors are then replaced by one capacitor that is rated for the full voltage. The four diodes are the full-wave bridge rectifier mentioned earlier. The transformer is connected to bridge terminals marked ac, and the other two terminals are marked plus and minus. When a full-bridge rectifier is used in this configuration, the current must flow through two diodes to charge the capacitor on each half-cycle.

The Voltage Doubler

The circuit in Figure 2.16 shows that a series diode and capacitor rectify the voltage and a dc voltage appears across the capacitor. If the positions of the capacitor and the diode are reversed, a dc voltage will still appear across the capacitor. One side of the capacitor is connected to the ac voltage source. The other side of the capacitor has this same ac voltage plus a dc value. If a second capacitor and diode are added, the result is rectification of the ac voltage superposed on the dc output of the first diode and capacitor. The resulting voltage is double the first dc voltage. This circuit is shown in Figure 2.20.

It is possible, through a chain of diodes and capacitors, to multiply the voltage by a higher factor. The last diode and capacitor place the

---

⚡ LEARNING CIRCUIT 13 ⚡
Building a Voltage Doubler

You will need (in addition to your circuit board and measuring equipment):

3 1.0-µF capacitors

2 1N1002 diodes

1 100-kΩ resistor

Build the voltage doubler circuit shown in Figure 2.20 (schematic) and Figure 2.21 (construction diagram.) Measure the output voltage using the oscilloscope or voltmeter.

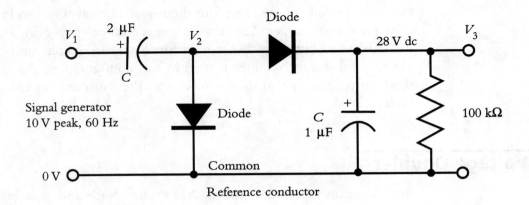

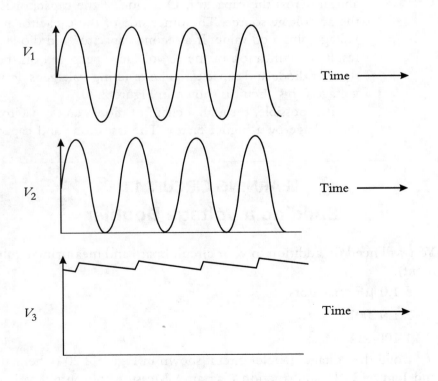

Figure 2.20 A voltage doubler circuit

74

Inductors, Transformers, and Resonance

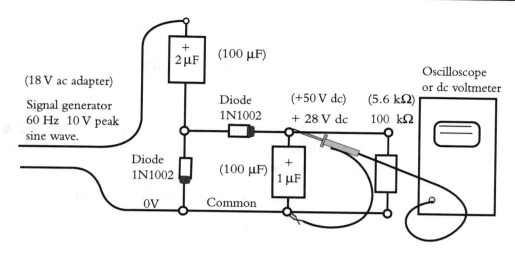

Observe the polarity markings on the capacitors. Notice the bar on the diode. The component values and operating voltages in parentheses assume the use of an ac voltage adapter. With the adapter, you can use lower values of load resistor.

The 2-µF capacitor can be made from two parallel 1-µF capacitors.

Figure 2.21 The construction of a voltage doubler circuit

capacitor to common. The first capacitor must be double the value of the second capacitor. The second diode must be able to handle a larger reverse voltage. Doublers and triplers are not very power-efficient, but if low current is required and the frequency is high enough the circuit can be very effective.

The circuit in Figure 2.20 requires dc to flow in the voltage source. If the current requirement is small, the transformer supplying this current will not saturate.

SELF-TEST

1. A dc voltage of 5 V across an inductor causes the current to rise at 1,000 A/s. What is the inductance?

2. 10 V dc is impressed across a 2-mH inductor. How long does it take for the current to reach 100 mA?

3. A sinusoidal current is 3 A peak at 60 Hz. What is the maximum rate of change of current?

4. What is the reactance of a 2-mH inductor at 50 kHz?

5. What is the reactance of a 50-µH inductor at 100 kHz?

6. Inductors of 1 mH, 2 mH, 3 mH, and 4 mH can be switched to provide all the values from 1 mH to 10 mH. Is there another group of four inductors you can use?

7. What is the parallel inductance of 1 mH and 500 µH?

8. What is the time constant for an inductance of 20 mH in series with a resistance of 10 kΩ?

9. A relay coil has an inductance of 2 H and a coil resistance of 200 Ω. How long does it take for the current to rise to 63% of final value?

10. A resistor of 100 Ω is in series with an inductor of 10 mH. What is the impedance at 1.592 kHz?

11. If the source voltage in problem 8 is 10 V, what is the current in one time constant?

12. In problem 10, what is the phase angle between the source voltage and the current?

13. In problem 8, what is the impedance at dc?

14. A 100-Ω resistor, a 10-mH inductor, and a 0.01-µF capacitor are in series. What is the resonant frequency?

15. In problem 14, what is the reactance of the capacitor at this frequency?

16. In problem 15, what is the reactance of the inductor at one-half the natural frequency?

17. In problem 15, what is the reactance of the capacitor at one-half the resonant frequency?

18. What is the impedance of the circuit in problem 14 at resonance?

19. What is the impedance of the circuit in problem 14 at half the resonant frequency?

20. What is the impedance of the circuit in problem 14 at twice the resonant frequency?

21. A 1-mH inductor and a 500-pF capacitor form a parallel resonant circuit. What is the resonant frequency?

22. In problem 21, 10 V is impressed on the circuit at resonance. What is the current flow in the inductor?

23. In problem 21, the resistance of the inductor is 20 Ω. What is the impedance of the circuit?

24. The open circuit voltages on two coils of a 120-V transformer are 25 V and 8 V. What are the turns ratios?

25. The load resistors in problem 24 are 50 Ω and 16 Ω. What is the primary current? Neglect magnetizing current.

26. The magnetizing current in a transformer is 50 mA. The primary voltage is 120 V at 60 Hz. The coil resistance is 30 Ω. What is the power dissipated?

27. In problem 26 the magnetizing current flows in a magnetizing inductance. What is the value of this inductance?

28. In problem 26, what is the input impedance?

29. A transformer secondary is marked 15-0-15 V. What are the two dc voltages that can be supplied?

30. A full-wave bridge is placed across the full voltage in problem 29. What is the output dc filtered voltage?

31. A 10-V dc voltage sags 0.5 V in each half-cycle of 60 Hz. The current averages 100 mA. What is the size of the filter capacitor?

32. A 10-V square-wave ac source is rectified. The frequency is 50 kHz. The load resistor is 200 Ω. Current is supplied every half-cycle through a diode. If the voltage can sag 0.1 V, what is the value of the filter capacitor?

78 PRACTICAL ELECTRONICS

ANSWERS

1. $V = L \times 1{,}000$. $V = 5$. $L = 5$ mH.

2. $V = L \times I/t$. $10 = 0.002 \times 0.1/t$. $t = 20$ μs.

3. Maximum current $= 2\pi f I_P = 1{,}130$ A/s.

4. $X_L = 628$ Ω.

5. $X_L = 31.4$ Ω.

6. 1, 2, 4, 8.

7. ⅓ μH.

8. $L/R = 2 \times 10^{-6}$ s.

9. $L/R = 0.01$ s.

10. Both reactances are 100 ohms. Total is 141.4 Ω.

11. Maximum I is 10 mA. 63% is 6.7 mA.

12. 45°.

13. 10 kΩ.

14. 1.592 kHz.

15. 10,000 Ω.

16. 5,000 Ω.

17. 20,000 Ω.

18. 100 Ω.

19. 15 kΩ.

20. 15 kΩ.

21. 225 kHz.

22. $X_L = 1{,}413$ Ω. $I = 7.1$ mA.

23. Power equals 1.0 mW. At 10 V this is 100,000 Ω.

24. If the primary has 1,200 turns, the secondaries are 250 and 150 turns.

25. The power is 16.5 W. At 120 V this is 0.137 A.

Inductors, Transformers, and Resonance 79

26. $I^2R = 0.075$ W.

27. X_L is much greater than R, so $X_L = 120/0.05 = 2,400\ \Omega$. The inductance is 6.4 H.

28. $2,400\ \Omega$.

29. The peak value less the diode drop is ± 20.6 V.

30. For a full-wave bridge, the voltage is 41.2 V.

31. A half-cycle is 8 ms. $C = Q/V = 0.1 \times .008/0.5 = 1600\ \mu F$.

32. $I = 10V/200\ \Omega = 50$ mA. $Q = 0.05 \times 10^{-5} = 5 \times 10^{-7}$ coulombs. $C = Q/V = 5 \times 10^{-7}/0.1 = 5\ \mu F$.

3 Introduction to Semiconductors

Objectives

In this chapter you will learn:

- about the simplest semiconductors: diodes, zener diodes, and transistors
- how diodes and zener diodes are used to clamp and limit signals
- how transistors are used as emitter followers
- the use of transistors to provide voltage gain

In this chapter you will expand your knowledge of electronic components to include semiconductors. In the 1950s, semiconductors were still an obscure subject known mainly to Ph.D.s working in what was called "solid state physics." In the 1960s they came out of the laboratory and into the marketplace, and today semiconductors are used in almost every electronic device we have. Their importance in electronics can hardly be missed.

The semiconductor components we will learn to use in this chapter include the diode, the zener diode, and the junction transistor. The diode we used in Learning Circuits 12 and 13 is the simplest of the semiconductor devices. Transistors are slightly more complex. In chapter 5 we will study integrated circuits, or ICs. An IC is a single component consisting of hundreds or perhaps thousands of interconnected transistors and resistors. Integrated circuits perform such tasks as amplification, timing, and waveform generation. Digital integrated circuits provide us with entire computers, microprocessors, and memories. Before we can understand ICs, we need to understand the way semiconductors work.

What Is a Semiconductor?

The first step in understanding semiconductors is to understand the basic materials from which they are made. The most common semiconducting materials are silicon and germanium. They are both basic chemical elements, and they are called *semiconductors* because they are halfway between being conductors and being insulators. In most applications, silicon is used rather than germanium because it can withstand higher operating temperatures. For this reason we will confine our discussion to silicon. But there are a few areas in electronics where germanium is used.

Silicon dioxide (the element silicon combined with oxygen) is the mineral quartz, which makes up most of the rock and sand on Earth. Pure silicon is rare in nature but can be grown as a crystal. A pure silicon crystal is an insulator. But when a very small percentage of impurity atoms are added to the crystal, it becomes a conductor. This is why it is called a semiconductor.

When phosphorous is added to the silicon crystal structure, a phosphorous atom replaces a silicon atom. The phosphorous provides an extra electron that is not needed in the bond that holds the crystal together. This electron is therefore free to move in the crystal, making the material a conductor. This material is called *n-type silicon*. The "n" refers to the free negative charge of the extra electron. The added phosphorous is called a *dopant* and the silicon is said to be *doped*.

When boron is added as a dopant, the silicon is called *p-type,* as the boron atom provides spaces (holes) that electrons can use to move through the crystal. P-type silicon is also a conductor.

Diodes

The diodes we used in chapter 2 were sandwiches of p- and n-doped silicon. In a p-n diode, electrical pressure can move free electrons from the n region to the p region. This is called the *forward direction of the diode*. In the other direction, electrons cannot move, as there are no electrons in the p region to move into the n region. This is called the *reverse direction*.

By convention, the direction of current flow is opposite to the direction of electron flow. So the direction of current flow is from the p to the n region of the diode. This direction is indicated by the direction of the arrow on the diode symbol; the arrow points toward the n material or *cathode* of the diode.

This limiting of current to one direction is the basis of diode action, which is also called *rectification*. Figure 3.1 shows the voltage/current curve for a typical silicon diode. Notice that a forward voltage of about 0.6 V is required before current flows.

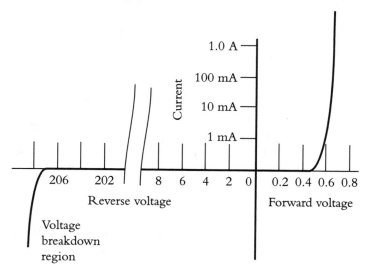

Note: The horizontal axis has two scales, and is broken on the left side to show the range. The breakdown voltage will vary between diodes. The diode is not expected to operate in this breakdown region.

Figure 3.1 The current/voltage curve for a silicon diode

84 PRACTICAL ELECTRONICS

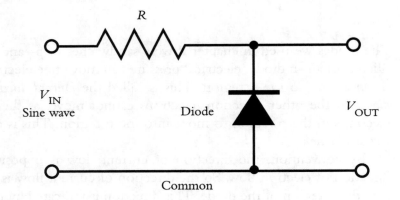

Typically the resistor R would be 1,000 Ω. This resistor value depends on the current required by the circuit connected across the output. If the output circuit is a load of 100 kΩ, the positive input voltage will be attenuated by 1%.

A typical power diode will provide clamping at frequencies below 20 kHz. For higher frequencies, a faster diode may be required.

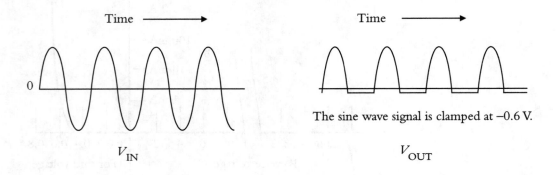

The sine wave signal is clamped at −0.6 V.

Figure 3.2 A diode clamp for negative voltages

The Diode as a Clamp

One application of a diode is a voltage clamp. The diode acts like a switch contact when it is conducting in the forward direction. The diode can act to clamp a signal so that it cannot change. When a voltage is clamped, the diode voltage drop is approximately 0.6 V whether

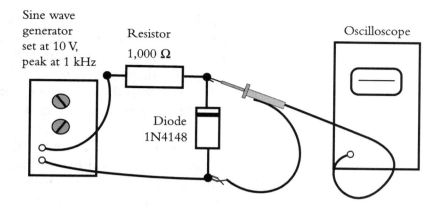

Figure 3.3 The construction of the circuit in Figure 3.2

the current is 1 mA or 1 A. In the reverse direction there is no current flow for a voltage difference of 1, 10, or 100 V. This is the characteristic of an open circuit. Figures 3.2 and 3.3 show a circuit that allows voltages greater than −0.6 V to pass and voltages less than −0.6 V to be blocked (clamped).

LEARNING CIRCUIT 14
Observing How a Diode Is Used to Clamp a Sine Wave

You will need (in addition to your circuit board and measuring equipment):

1 1-kΩ resistor

1 1N4148 diode

1. Connect the diode and the resistor as shown in Figures 3.2 and 3.3 to the sine wave output of a function generator.
2. Use the oscilloscope to observe that the sine wave cannot go negative. In other words, the negative voltage is clamped.
3. Reverse the diode and note that the positive voltage is clamped.

Note that from this point forward in the book, when two consecutive figures show a Learning Circuit, the schematic drawing will be given first and the construction diagram will follow.

In this circuit, the diode conducts only when the signal is more negative than 0.6 V relative to the common or reference conductor. If the diode is reversed in direction, the circuit allows only voltages less than +0.6 V to pass.

When a diode clamps a voltage source, it acts as a short circuit. This short circuit can cause overload or overheating if the source is a low impedance. In this case a resistor is usually placed between the voltage source and the diode. If the maximum voltage that can be tolerated is 10 V and the maximum current is 10 mA, the resistor must be 1,000 Ω or greater.

Diodes are often used to limit a signal voltage so it cannot exceed the power supply voltages. This is done to protect the circuit against damage. In this case the diode clamps are connected between the signal and the power supply voltages. This circuit is shown in Figures 3.4 and 3.5. If the signal voltage is between the limits of the power supply voltages, the signal is unaffected (unclamped).

LEARNING CIRCUIT 15
Diode Clamping to a Power Supply

You will need (in addition to your circuit board and measuring equipment):

   2 9-V batteries

   2 1N4148 diodes

   1 1-kΩ resistor

Connect the circuit in Figures 3.4 and 3.5. Use a sine wave generator to show that voltages above 9.6 V are clamped.

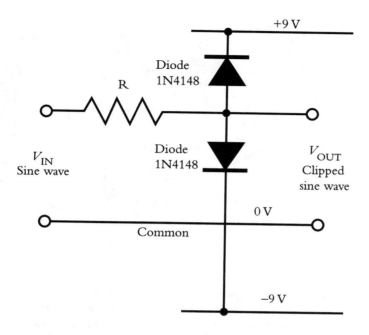

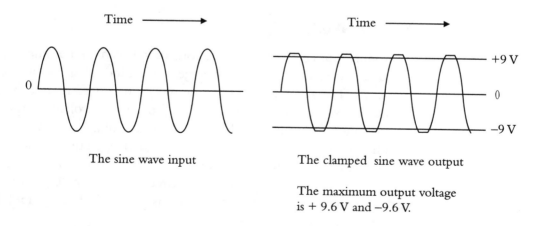

Figure 3.4 Diode clamps used to block signal voltages greater than the power supply voltages

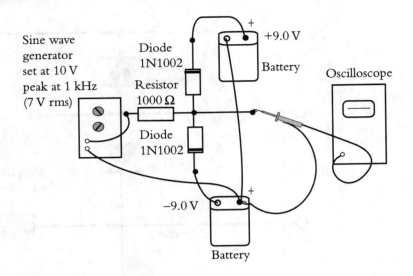

Note: The only electrical connections are at terminal points indicated by the dots. If bus wire is used for connections, the crossings should be spaced apart. In a more permanent circuit, the connections may be sleeved (insulated) to avoid a wire being bent, with a resulting short circuit.

Figure 3.5 The construction of the circuit in Figure 3.4

The Zener Diode

By proper doping of the pn junction, diodes can be designed with a controlled reverse breakdown voltage. This kind of diode is called a *zener diode.* The reverse voltage rating for a diode used in a power supply is often 400 V. In a zener diode the breakdown voltage is specified and can range from 3.4 V to 100 V. The zener voltages in common use range from 5 V to 20 V. In the normal range of operation, zener diodes are operated in the breakdown or reverse direction. Both diodes and zener diodes conduct in the forward direction. Zener diodes begin to conduct when the breakdown voltage is reached. The symbol for a zener diode is shown in Figure 3.6.

Zener diodes are typically ¼ W components. A 10-V zener with 20 mA reverse current dissipates 0.2 W. The voltage across a zener diode varies over a narrow range as the current changes. The circuit symbol for a zener diode is ⟊. The arrow is in the direction of zener current flow. A typical zener diode voltage/current curve is shown in Figure 3.6.

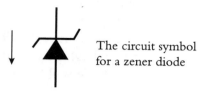

The circuit symbol for a zener diode

The arrow is in the direction of zener current flow.

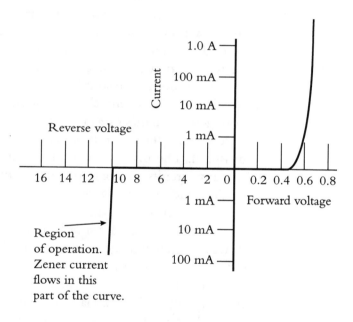

Note: The horizontal axis has two scales.

Figure 3.6 The voltage/current curve for a 10-V zener diode

The flow of reverse current is the result of an avalanche effect. When a small number of electrons start to flow, they knock loose many more and start the avalanche. The ideal zener would have a very sharp transition between an open circuit and the point where current flows. This transition region is called the *knee of the curve.* In general, the knee is not very sharp for zener diodes rated less than 4 V.

Zener diode voltages do change with temperature. The most stable range is around 5.6 V. Zener diodes in this voltage range are often used as voltage references because of their stability.

Zener diodes should never be placed directly across a power supply. If the voltage exceeds the breakdown voltage, there is a chance that the zener diode will be destroyed or the power supply will be damaged.

Source Impedance

In chapter 1 we looked at the internal resistance of a battery. We also need to know how to make this measurement for a zener diode. When we know how to do this, we will be able to apply the same measuring method to circuits. The circuit needs to be active or the component needs to be drawing current before the measurement makes sense.

By definition, the source impedance of a circuit is the ratio of voltage change to current change for sine waves. The term *impedance* is more general than *resistance,* as it can include the reactance of an inductance or capacitance. In most situations the resistance is all that is being measured. The internal resistance of a battery is its source impedance. In many circuits that provide an output voltage, the source impedance is called the *output impedance* or simply the *output resistance.* All these expressions mean the same thing. If the measure is not made using sine waves, then the term *resistance* rather than *impedance* should be used. However, the language of electronics is not always perfectly consistent, and this is a good example.

When the source resistance of a circuit is measured, the changes must be within the normal range of operation. For example, an automobile battery has a very wide range of operation. If a load resistor of 1.2 Ω is applied across a 12-V automobile battery, the voltage might drop to 11.8 V. The change of voltage is 0.2 V. The change of current is 10 A. The battery has an output resistance of 0.2/10 = 0.02 Ω. But for electronic power sources, a 10-A current flow might be outside the normal range of operation.

Output impedance can be measured dynamically by driving the circuit from a sine wave signal generator using a series resistor. A coupling capacitor should be used if there is an offset voltage. Assume the generator voltage is 10 V and the resistor is 1,000 Ω. The resistor and the output impedance form a voltage divider. The voltage at the output might be 0.1 V. The ratio of 10 V to 0.1 V is an attenuation factor of 100. The output resistance is 1,000/100 = 10 Ω. These voltages can best be observed on an oscilloscope. The waveforms must both be sine waves or the measurement is invalid. The reactance of the capacitor must be small compared to 1,000 Ω to make the measurement valid.

The Zener Diode as a Voltage Regulator

The voltage across a zener diode is nearly constant over a wide range of current. This makes a zener diode useful as a voltage regulator. Figures 3.7 and 3.8 show an application where a source voltage that varies from 9 to

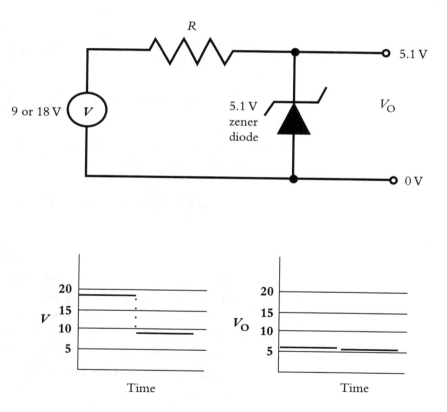

Note: The current flowing in R supplies both zener current and load current. If the load current is 5 mA, the zener current is 10 mA, and the voltage source 15V, the resistor value is 1,200 Ω.

Figure 3.7 A zener diode voltage regulator

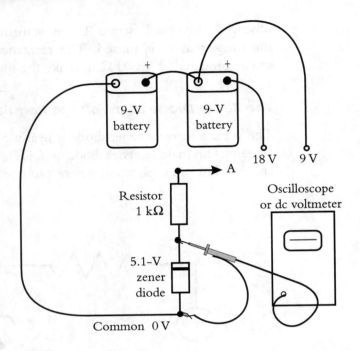

Connect point A to 9 V or 18 V to illustrate that the voltage across the zener diode remains nearly constant.

Figure 3.8 The construction of the circuit in Figure 3.7

18 V provides current for a 5.1-V zener diode. The voltage across the zener diode varies less than 0.05 V. The zener voltage changes only 1% for a 100% change in source voltage. This is a regulation factor of 100.

The source impedance of this zener voltage source can be determined by placing a load resistor across the 5.1-V zener diode. If the load is 1,000 Ω and the voltage changes 0.03 V, the output impedance is $0.03/0.005 = 6$ Ω.

When a zener diode is used as a power supply, any added load current is taken from the zener diode current. Assume that in Figure 3.7 the unloaded zener diode current is 10 mA. If the added load resistor is 1,000 Ω, the zener current drops to 5 mA.

A zener diode can be used as a clamp so that a signal cannot exceed the zener voltage. In the negative direction the forward direction of the

Introduction to Semiconductors 93

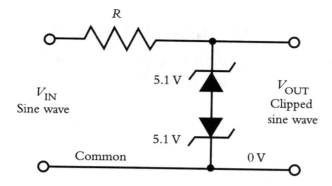

These zener diodes are connected back to back.

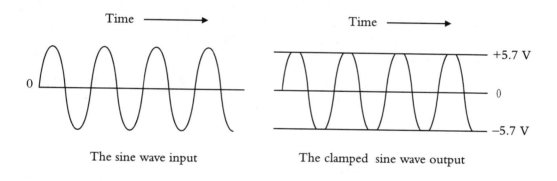

The sine wave input   The clamped sine wave output

The clipped voltage is the zener voltage plus one diode voltage drop. When one zener diode is in the zener region, the other diode is forward-biased.

A resistance R must be present or the circuit cannot function. Sometimes this resistance is a part of the signal source.

Figure 3.9 Two zener diodes acting as a signal clamp

zener clamps the signal as a standard diode. To limit the voltage symmetrically, two zener diodes can be used back to back. This circuit is shown in Figure 3.9. If the zener voltages are 5.1 V, the clamping voltage is actually 5.7 V (the zener voltage plus a diode drop of 0.6 V). In this example the signal voltage cannot exceed ±5.7 V.

LEARNING CIRCUIT 16
Observing Voltage Regulation Using a Zener Diode

You will need (in addition to your circuit board and measuring equipment):

    2 9-V batteries

    1 1-kΩ resistor

    1 5.1-V zener diode

1. Connect the circuit shown in Figures 3.7 and 3.8, using two 9-V batteries in series to supply current to a 5.1-V zener diode.
2. Measure the voltage across the zener diode.
3. Now use one of the batteries and note the zener voltage. What was the regulation factor? The bar on the diode corresponds to the bar in the schematic.

LEARNING CIRCUIT 17
Observing Voltage Clamping with Two Zener Diodes

You will need (in addition to your circuit board and measuring equipment):

    1 1-kΩ resistor

    2 5.1-V zener diodes

Construct the circuit shown in Figures 3.9 and 3.10, placing two 5.1-V zener diodes back to back along with a resistor across a sine wave voltage source. Observe that the signal voltage is limited by the zener voltage to ±5.7 V.

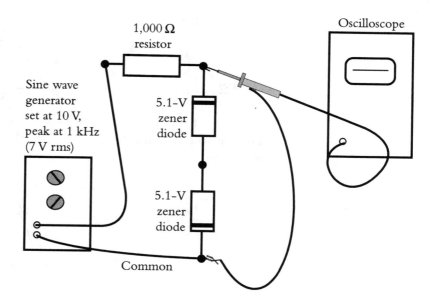

Figure 3.10 The construction of the circuit in Figure 3.9

The NPN Transistor

We will now take up the subject of transistors. Transistor are called *active elements,* as opposed to resistors, capacitors, and inductors, which are *passive elements.* Passive elements simply react to a signal without enhancing it, but active elements like transistors use input from a power supply to enhance a signal in some way. Diodes and zener diodes are considered nonlinear passive elements, which means that they change the waveform but do not otherwise add to it.

Figure 3.11 shows an npn transistor. This transistor is formed by placing p-type material between two layers of n material.

The outer n layers are called the *emitter* and the *collector.* The p material is called the *base.* The two n layers may appear to be symmetrical, but they are not; they must be connected in the correct order as specified by the manufacturer. If you interchange the emitter and the collector, you may damage the transistor. If the base is disconnected, the pnp layers form two back-to-back diodes, which means that current is theoretically blocked and cannot flow in either direction.

If current is caused to flow in the forward direction in the base/emitter junction, the p material is supplied with electrons and the

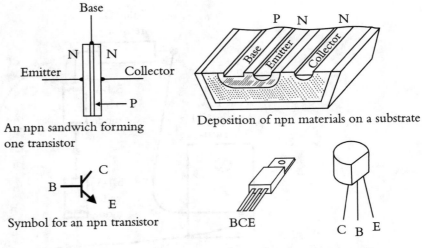

Figure 3.11 An npn transistor and its circuit symbol

npn stack takes on the character of all n material. If there is a collector voltage, a small flow of base current will control the current that flows from the emitter to the collector. This control of collector current is the essence of the transistor.

In a transistor, current flows in the forward direction in the base/emitter junction. The collector current flows in the reverse direction through the collector/base junction. When there is no base current, the collector/base junction is a diode in the reverse direction. This junction must be designed to withstand a reverse voltage equal to the power supply voltage, while the base emitter junction does not ordinarily see a significant reverse voltage. In most transistor designs the base/emitter reverse voltage is limited to a few volts. This is the reason the emitter and the collector cannot be interchanged in function.

The relationship between base current, collector current, and collector voltage in a typical npn transistor is shown in Figure 3.12. The individual curves represent a fixed base current. Notice that the collector current is controlled by the base current and not by the collector voltage.

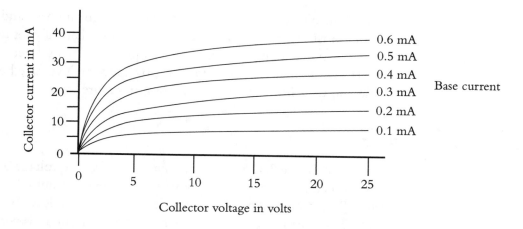

Note: The base current is fixed along any one curve.

Figure 3.12 The operating curves for a typical npn transistor

Transistor Gain

Transistors that operate along one of the curves in Figure 3.12 are operating in their linear range. This is the operating range we will consider in this section. There is a distinction between operating voltages and currents and signal voltages and currents. A signal makes changes to the base current and voltage. These changes result in new collector voltages and currents. The ratio of changes to two parameters is called *gain*.

The ratio of a change of base current to the change in collector current is called the β (beta) of the transistor. This is current gain. As an example, if the base current changes from 1.0 mA to 1.01 mA and the collector current changes from 10 to 11 mA, the value of β is 100. The distinction between signal voltage and operating voltage is clear when the signal is a small ac voltage superposed on the operating voltages.

Another measure of gain is the ratio of base voltage change to collector current change. This is called *transconductance*. This measure of gain is often given in transistor specifications. The higher the transconductance, the more voltage gain the transistor can provide.

In most circuits it is desirable to have a voltage gain, not a current gain. The transistor must be a part of a circuit so that the current gain (β) of a transistor can provide this voltage gain. Changes in collector current can be converted to changes in voltage by using a collector resistor. Controlling the voltage difference between the base and emitter is not as easy. To be effective, a circuit must automatically set this difference voltage so that the transistor is operating in its linear range. Attempts to obtain gain from a transistor by applying a voltage between the base and the emitter will usually result in failure.

The Emitter Follower

The circuit in Figure 3.13 is called an *emitter follower*. The voltage between the base and emitter is automatically set to provide the required collector current. When the base moves one volt, the emitter follows by almost one volt. This means the gain of the emitter follower circuit is very close to unity. This is a good time to define the term *voltage gain*. Voltage gain is the ratio of output voltage change to input voltage change.

In the circuit of Figure 3.13, assume the input base voltage is set to

---

LEARNING CIRCUIT 18
Constructing an Emitter Follower

You will need (in addition to your circuit board and test equipment):

1 2N3904 (npn) transistor

1 2.7-kΩ resistor

For this circuit, use the dc power supply you built in Learning Circuit 12. Add the emitter follower shown in Figures 3.13 and 3.14.

The current in the transistor is controlled by the base voltage, the power supply voltage, and the emitter resistor. If the base is at 0 V, the emitter is at about −0.6 V. The voltage drop across the emitter resistor is the negative power supply voltage minus 0.6 V. If the signal voltage is +10 V and the resistor is 2,700 Ω, the current is 9.4/2,700 = 3.5 mA. If the β of the transistor is 200, the base current is 17.4 µA. This signal voltage level can be provided by an external battery, or you can use a square wave function generator at a low frequency. Use the oscilloscope to verify that the emitter voltage follows the base voltage except for a slight offset.

Leave this circuit connected to the power supply, as you will also need it for the next Learning Circuit.

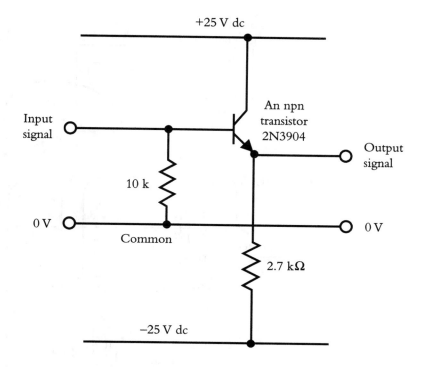

The 10-kΩ resistor provides a path for base current when the input signal is disconnected or the input signal does not provide a path for this current. Base current must flow or the transistor cannot function.

Assume there is no input signal connected. The emitter current is about 10 mA. It is determined by the emitter resistor and the minus power supply voltage. If the β of the transistor is 100, the base current is 100 μA. This current flows in the 10 kΩ resistor and the base voltage rises to 1 V. This means that the emitter voltage is actually about 0.4 V.

If the input signal lead is connected to common through a low value of resistance, the base is at 0 V. This means that the emitter voltage is −0.6 V.

The output voltage follows the input voltage except for an offset of 0.6 V.

Figure 3.13 An emitter follower circuit using a dual voltage power supply

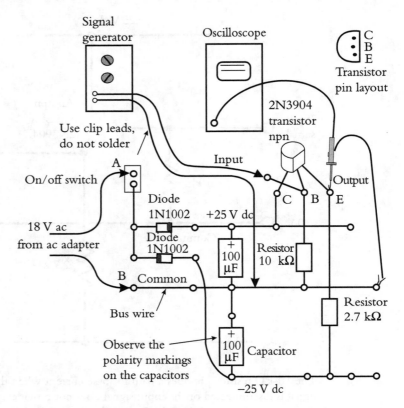

Figure 3.14 The construction of the circuit in Figure 3.13

+2 V. The emitter voltage follows and is at +1.4 V. The new emitter current is 11.4/2,700 = 4.2 mA. The new base current is 21.1 µA. This means that the base has moved 1 V and the base current has changed 3.7 µA. The ratio between voltage change and current change is called *input impedance*.

In the circuit shown in Figure 3.13, the ratio of changes is $V/I = 1/3.7 \times 10^{-6}\ \Omega \cong 270{,}000\ \Omega$. (This measure assumes the 10-kΩ resistor is not present.) If the transistor β is 400, the input impedance would be 540,000 Ω. The input resistance of a transistor that is not used as an emitter follower is typically around 100 Ω. (Note that this is an example of when you are likely to hear the term *impedance* used in place of *resistance*. Impedance is incorrect, since sine waves were not used in the measurement. However, in this circuit, measurement using sine wave voltages would give the same result.)

In Figure 3.15, assume the base voltage has been adjusted so that the emitter voltage is +5 V. A 100-Ω load resistor R is placed between this emitter and the common lead. The current in the resistor must flow in the transistor. This means the base current must increase slightly. The emitter voltage might drop 0.05 V to 4.95 V.

In this example the voltage change is 0.05 V and the current change is 50 mA. The output impedance is $V/I = 1$ Ω. This example shows that an emitter follower has a low output impedance.

Emitter followers have a gain very close to 1, a high input impedance, and a low output impedance. In effect, an emitter follower is an impedance converter. A voltage source with a high source impedance or with a

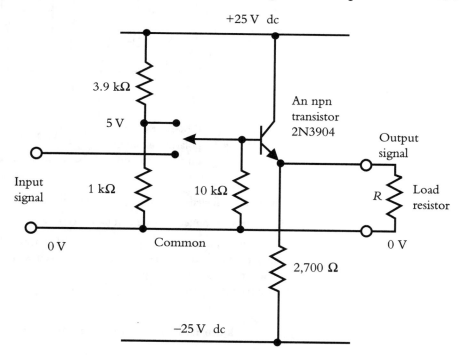

This circuit shows two signal sources. The first is 5 V dc and the second is a voltage from a signal generator.

When the input signal goes positive, the emitter follows. The current in the load resistor flows in the transistor and adds to the static current. If the input signal goes negative, the current in the load is subtracted from the transistor current. The negative output voltage is limited by the static current in the transistor. This limits the negative voltage swing of the emitter follower.

Figure 3.15 An emitter follower with a load resistor

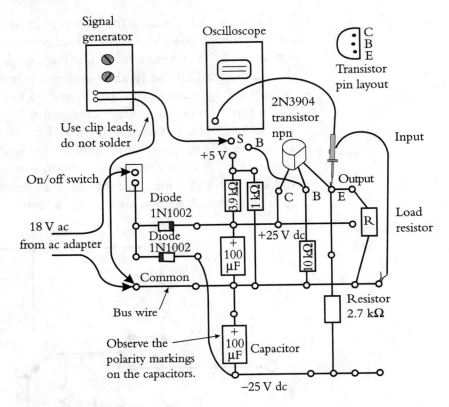

Figure 3.16 The construction of the circuit in Figure 3.15

The resistor labeled R is the output load resistor.

The transistor connections are as follows: C = collector, B = base, and E = emitter

To observe an output signal, jumper the signal generator S to the base B. Jumper B to 5 V to bias the input positive. If you want to have a bias and a signal at the same time, the signal generator must be coupled through a capacitor. A recommended capacitor value might be 0.1 μF.

limited ability to supply current can be converted to a voltage source with a low source impedance that can supply current. In our example, the input current is in the microampere range and the output current is in the milliampere range. This multiplication of current is one form of gain.

The emitter follower follows an input signal no matter how slowly it changes. Amplifiers that handle this type of signal are called *dc amplifiers*. An emitter follower is a dc amplifier with a gain of 1. If the input

LEARNING CIRCUIT 19
Observing the Output Impedance of an Emitter Follower

You will need (in addition to your circuit board and test equipment):
  1 100-Ω, 1 2.7-kΩ, 1 1.0-kΩ, and 1 3.9-kΩ resistor

1. Use your power supply and the emitter follower you built in Learning Circuit 18.
2. Add a voltage divider to set the input base voltage to +5 V as shown in Figures 3.15 and 3.16. The voltage divider can be determined as follows: If the power supply voltage is 25 V, set the divider current to 5 mA. The top resistor is 4 kΩ (we use 3.9 kΩ, as it is the nearest standard value), and the bottom resistor is 1 kΩ.
3. Measure the emitter voltage with respect to the common lead. It should read about 4.4 V.
4. Place a 100-Ω resistor from the emitter to common. This is resistor R in Figure 3.16.
5. Measure the emitter voltage. What is the output impedance? Save this circuit construction.

LEARNING CIRCUIT 20
Measuring the Gain of an Emitter Follower

You will need:
  The circuit from Learning Circuit 19

1. Use the circuit you built in Learning Circuit 19. Measure the input base voltage and the emitter output voltage with respect to power supply common.
2. Use a clip lead to connect the base to the power supply common (this removes the 5-V input base signal) and measure the emitter voltage. Use these voltage measurements to determine the gain.
3. Remove the 100-Ω resistor load from the emitter to the common.
4. Repeat these steps to obtain new voltages. What is the new gain?

voltage changes 1 V at dc, the output voltage changes 1 V at dc. The 0.6-V voltage difference is an offset signal. In applications such as audio amplifiers where only ac is of interest, a capacitor can block the offset voltage. This capacitor and any terminating resistor form a high-pass filter. The −3 dB point occurs at the frequency where the reactance of the capacitor equals the terminating resistor.

The emitter follower in Figure 3.15 can supply current to a load resistor. When the input voltage is positive, the output voltage follows. The load current is added to the collector current. In the negative voltage direction, the current in the load is subtracted from the collector current. In the negative direction the emitter follower cannot support a load current that exceeds the nominal collector current. The collector current for various output voltages is shown in Figure 3.17. In the next chapter we will discuss a circuit that does not have this limitation.

The emitter follower circuit using a single voltage power supply is shown in Figure 3.18. A voltage divider must be used to establish a base voltage so that the transistor can draw current. This transistor current is determined by the base voltage and the value of the emitter resistor. The voltage divider must supply any required base current.

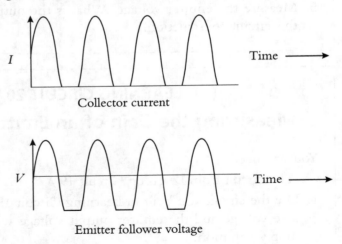

The collector current is the static current plus the load current. When the output voltage is negative, the load current is subtracted from the static current. When the load current takes all the collector current, the emitter follower cuts off.

Figure 3.17 The current and voltage in an emitter follower

It is a good idea for the voltage divider current in Figure 3.18 to be 10 times the base current. If the base current is 100 μA, the divider current can be 1 mA. Two coupling capacitors are required if the operating voltages are to be blocked. In this circuit the input impedance is dominated by the voltage divider. This impedance is equal to the two divider resistors in parallel. The value of the output capacitor depends on the terminating resistor, not on the source impedance.

You will see in chapter 5 that an integrated circuit can be used to build the equivalent of an emitter follower. This approach is often less

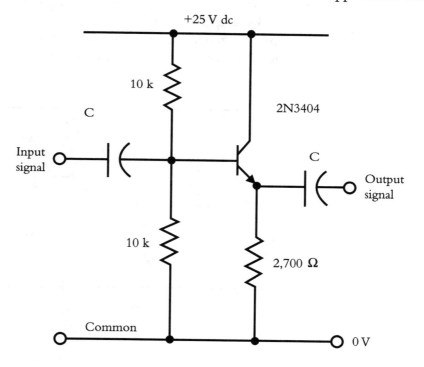

The coupling capacitors block the operating voltages. The emitter follower functions at ac frequencies determined by the RC cut-off frequencies. On the input the input resistance is 5 kΩ. The −3dB point is calculated at the frequency where the reactance of the capacitor equals 5 kΩ.

The output capacitor value depends on the load resistance. If the −3dB frequency is to be the same as the input, then the reactance of the capacitor equals the output load resistance (not shown) at this frequency.

Figure 3.18 An emitter follower and a single voltage power supply

expensive and has the additional advantage of having no offset problems. Emitter followers are still used when the output current must be greater than that supplied by an IC amplifier, for example in supplying current to a loudspeaker.

Voltage Gain Using an NPN Transistor

The circuit in Figure 3.19 provides a negative gain of 3.0. In this circuit we make use of a dual 25-V power supply, two zener diodes, resistor $R_1 = 3.3$ kΩ, and resistor $R_2 = 10,000$ Ω.

The two zener diodes are used to provide a -5.1-V power supply voltage. The transistor current can vary, but the zener voltage stays con-

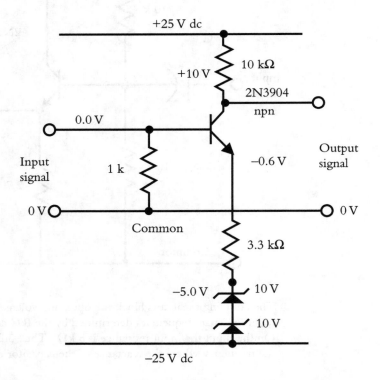

The output signal is biased about 10 V positive. In an ac application this voltage can be blocked by a capacitor.

Figure 3.19 A voltage gain using an npn transistor

stant. When the base voltage is 0, the emitter voltage is approximately −0.6 V. The voltage across the emitter resistor is 4.4 V. The current in the emitter resistor is 4.4 V divided by R_1 = 1.333 mA. This current flows in the collector resistor R_2. The voltage drop in this resistor is IR_2 = 13.33 V. If the power supply voltage is 25 V, the collector voltage is 11.7 V. These are operating voltages.

If the input voltage rises 1.0 V, the new emitter voltage is +0.4 V. The current in the emitter resistor increases to 1.8 mA. This current flows in the collector resistor. The resulting voltage drop is 18 V. The new collector voltage is 8.7 V. The collector voltage dropped 3 V. The gain is the ratio of output voltage change to input voltage change, or −3 V/1.0 V = −3.00. This is also the ratio of R_2/R_1. If the base voltage changes sinusoidally, the output voltage also changes sinusoidally. When the sine wave has a peak value of +1.0 V, the output signal has a peak voltage of −3 V.

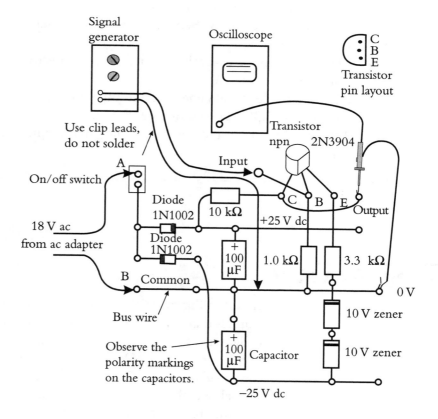

Figure 3.20 The construction of the circuit in Figure 3.19

> ─⏦─ **LEARNING CIRCUIT 21** ─⏦─
>
> ## Obtaining Voltage Gain from a Transistor
>
> You will need (in addition to your circuit board and test equipment):
>
>   1 2N3904 transistor
>
>   1 1-kΩ, 1 3.3-kΩ, and 1 10-kΩ resistor
>
>   2 10-V zener diodes
>
> 1. Build the circuit shown in Figures 3.19 and 3.20, using the circuit values indicated. Be careful to connect the transistor correctly. A diagram of the transistor pin layout is shown in the upper right corner of Figure 3.20. Check that the dc voltages are correct.
> 2. Set the level of a sine wave signal generator to 1 V peak-to-peak at 1 kHz. Verify this on your oscilloscope.
> 3. Connect the generator between the transistor base and the power supply common.
> 4. Measure the peak-to-peak ac signal at the collector. Verify that the gain is −3.
> 5. Disconnect the signal generator and turn off the power.
> 6. Change the 3,300-Ω resistor to 2,200 Ω.
> 7. Turn on the power and verify that the circuit gain is now approximately 5.

Voltage Gain at AC

The circuit in Figure 3.19 can be modified to have a gain of 10 at ac but a gain of less than 1 at dc. This circuit is shown in Figures 3.21 and 3.22. The collector resistor is set to 10 kΩ. The emitter resistance is made up of two resistors: $R_1 = 1$ kΩ and $R_3 = 22$ kΩ.

The capacitor across R_3 has a reactance of 1 kΩ at 159 Hz. At frequencies above about 150 Hz the gain is the ratio of $R_2/R_1 = -10$. Below about 10 Hz the gain is less than 1. The circuit functions like a

Introduction to Semiconductors

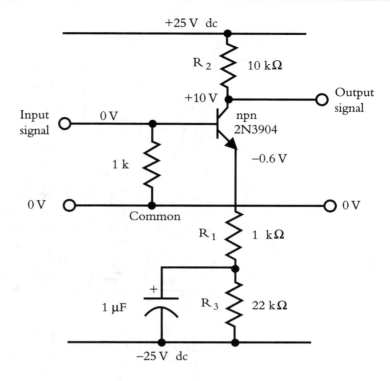

This circuit has a gain of 10 at frequencies above 10 Hz. The output point is biased at plus 10 V. An output capacitor is needed to block this offset.

Figure 3.21 An npn transistor circuit with a gain of 10 at ac

high-pass filter except that the gain never falls to 0. The capacitor value is about 1 μF.

In Figure 3.19 the zener diode held the emitter resistor return to −5 V at all frequencies. In Figure 3.21 a capacitor holds the emitter resistor return fixed at frequencies above 150 Hz. The advantage of this circuit is that the dc operating conditions do not vary, as there is no gain at dc. This technique is used in high-frequency amplifiers where dc gain is unimportant and stable operating conditions are desirable.

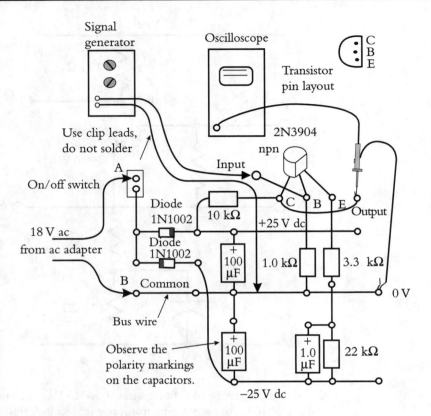

Figure 3.22 The construction of the circuit in Figure 3.21

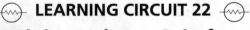

LEARNING CIRCUIT 22
Obtaining Voltage Gain from a Transistor at AC

You will need (in addition to your circuit board and test equipment):

1 2N3904 transistor

2 1-kΩ, 1 10-kΩ, and 1 22-kΩ resistor

1 1-μF capacitor

1. Build the circuit of Figures 3.21 and 3.22.
2. Set the sine wave generator output to 0.5 V.
3. Test the gain at 100 Hz, 10 Hz, and 1 Hz.
4. Remove the capacitor and note the gain at 150 Hz.

Introduction to Semiconductors

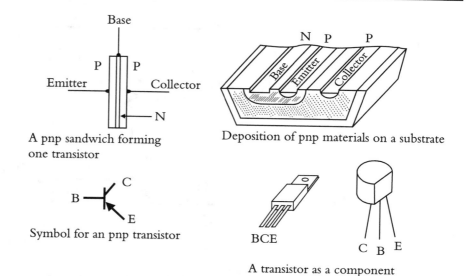

Figure 3.23 The structure and symbol for a pnp transistor

The PNP Transistor

Figure 3.23 shows the structure and symbol for a pnp transistor. The outer two layers of p-type silicon are called the *emitter* and the *collector* and the center n silicon layer is called the *base*. This pnp transistor is very similar to the npn transistor except that all the voltages are reversed. In the next section the pnp transistor is used as an emitter follower.

The PNP Emitter Follower

The emitter follower using the pnp transistor is the mirror image of the npn transistor. This circuit is shown in Figures 3.24 and 3.25.

Note that the offset voltage is +0.6 V instead of −0.6 V. In this circuit any output load current for a negative output voltage is added to the transistor current. In the positive direction the output load current subtracts from the collector current. When the transistor current is zero the output voltage is at its maximum.

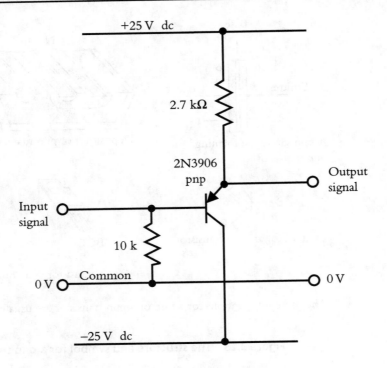

Figure 3.24 The pnp emitter follower

⌇ **LEARNING CIRCUIT 23** ⌇
Observing the Gain of a PNP Emitter Follower

You will need:

 The circuit you built in Learning Circuit 22

 1 2N3906 transistor

Using Learning Circuit 22, modify the circuit to that shown in Figures 3.24 and 3.25. Go through the same steps as before. The pnp and npn transistors have the same pin arrangement. Verify that the operating voltages are correct. Before connecting the signal generator, verify that the signal level is 1 V peak-to peak. Measure the ac gain with and without a load resistor of 100 Ω.

Introduction to Semiconductors 113

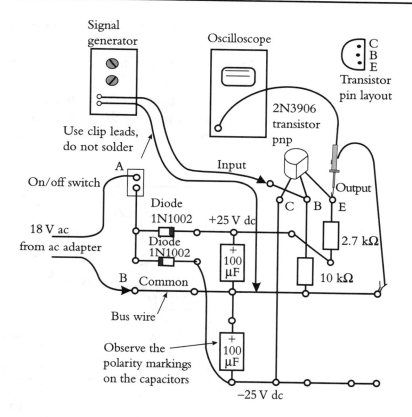

Figure 3.25 The construction of the circuit in Figure 3.24

Using the NPN and PNP Transistors Together

It is common practice in drawing circuits to draw the positive voltages as a horizontal line across the top of the schematic. Similarly, the negative voltage is a horizontal line at the bottom of the schematic. These parallel lines have reminded people of railroad tracks. This is probably the origin of the term *power supply rails*. The power supply rails are simply the conductors that carry the power supply voltages. Another term used to describe a power supply lead that runs through a circuit is a *bus*. The term probably started in the power industry where large power conductors are used. The term extends to ground or other common leads.

The convention of placing the positive power supply conductor at the top of the page means that current always appears to flow from top to bottom. The arrows on the emitters of npn and pnp transistors point in the direction of current flow and thus they always point toward the bottom of the page. These conventions make it easier to read a schematic.

An example of how npn and pnp transistors work together is shown in Figures 3.26 and 3.27. In this circuit each transistor provides a gain of 3.03 and the output collector is at the input signal common potential.

The base current for the pnp transistor is supplied by the collector of the npn transistor. If the β of the pnp transistor is 200, this base cur-

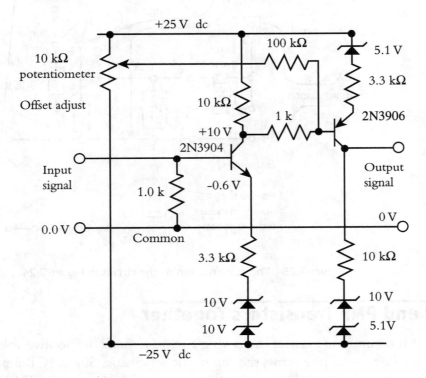

A potentiometer is a resistor with a contact that slides along the length of the resistance. This component provides a continuous voltage divider.

When the input is at 0 V the output may be offset by a small voltage. To correct for this offset, a small current can be injected into the base of the pnp transistor through 100 kΩ. The potentiometer allows this current to be adjusted.

This is a dc amplifier with a gain of 10. It amplifies signals from dc to over 100 kHz.

Figure 3.26 An npn and pnp transistor used to provide gain

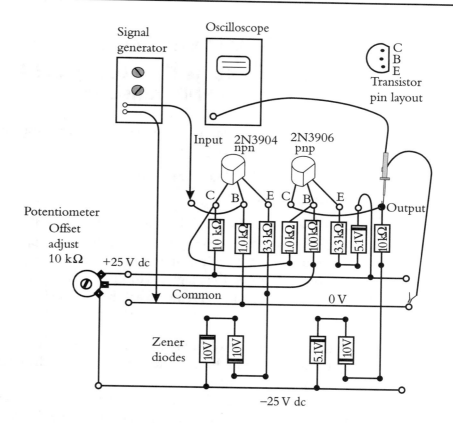

The power supply details are not shown in this diagram.

Figure 3.27 The construction of the circuit in Figure 3.26

rent is a small fraction of the collector current. The overall gain of this circuit is +9.2. If the first stage emitter resistor is shunted by 10 kΩ, the overall gain will be very nearly 10.

The circuit of Figure 3.26 may have an offset of several volts. This offset can be removed by injecting a small dc signal into the base of the second stage. A zeroing control is provided, as the amount of offset is unknown. The control component is called a *potentiometer*. It is a resistor with a slider that makes contact along the resistance. The potentiometer forms an adjustable voltage attenuator. In this circuit the slider provides any voltage from +25 V to −25 V. This variable voltage changes the base current to the second stage and corrects for the offset. Later, when we use an IC amplifier, this adjustment may not be necessary.

LEARNING CIRCUIT 24
Providing Voltage Gain Using an NPN and a PNP Transistor

You will need (in addition to your circuit board and test equipment):
  1 10-kΩ potentiometer
  1 2N3904 and 1 2N3906 transistor
  3 10-V and 2 5.1-V zener diodes
  2 10-kΩ, 2 1.0-kΩ, 2 3.3-kΩ, and 1 100-kΩ resistor

Note that on this Learning Circuit the power supply is no longer shown. You are now building circuits that are more complex, and there is not enough room on the diagrams to show it along with the new circuit. On future Learning Circuits it will be assumed that a power supply is present, but it will no longer be shown.

1. Build the circuit shown in Figures 3.26 and 3.27 on your power supply board. Be careful to measure the voltages on the circuit after the power has been turned on.
2. Set the sine wave signal to 0.2 V peak-to-peak. If this is difficult to do, attenuate the signal generator by using a voltage divider of 1,000 Ω and 100 Ω. Use the midpoint on the divider to provide a reduced signal.
3. Measure the gain of the circuit using the oscilloscope. If the offset is a problem, adjust the potentiometer to set the output voltage to 0.

Offsets and dc drift are always a problem with circuits that amplify both slowly and rapidly changing voltages. This is another example of a dc amplifier.

The circuit in Figure 3.26 is an example of obtaining voltage gain using transistors. This circuit is useful in learning about input and output impedances, gain, and zero shifting. The preferred way to obtain this gain is with an IC amplifier, which we will discuss in the next chapter. The circuits we have studied so far are a part of the inner workings of an IC amplifier. Understanding these circuits is a vital part of understanding electronics.

SELF-TEST

1. The current in a 15-V zener diode is 10 mA. What is the power dissipation?

2. If the load resistor across the zener diode in problem 1 is 3 kΩ, what is the power dissipation in the zener diode?

3. The voltage across a zener diode is 9.75 V. When a 5-kΩ resistor is placed across the diode, the voltage drops to 9.60 V. What is the source impedance?

4. The collector current in an npn transistor is 2 mA. The collector resistor is 2,000 Ω. If the base current is increased by 2 μA and the collector voltage changes 1 V, what is the β of the transistor?

5. The input to an emitter follower changes from −0.5 to 1.5 V. The emitter changes from −1.2 to 0.75 V. What is the gain of the emitter follower?

6. The input current in problem 5 changes from 0.8 mA to 0.9 mA. What is the input impedance?

7. An npn emitter follower supplies 12 V to a 200-Ω resistor. The voltage changes to 11.8 V when the second 200-Ω resistor is added. What is the output impedance?

8. In problem 7, if the collector current for no load is 20 mA, what is the collector current when both loads are present?

9. A pnp transistor is used as an emitter follower. The output voltage for no load is 30 mA. What is the maximum positive voltage that can be supplied to a 500-Ω load?

10. The power voltage varies from 105 V to 125 V. An emitter follower requires a minimum of 5 V dc from the collector to the emitter. A dual power supply is used to supply voltages for an emitter follower. The emitter follower is used to regulate 15 V dc. Allow for one diode drop and a 2-V drop in the transformer coil plus a peak-to-peak ripple voltage of 1 V. What should the unloaded rms voltage be for the secondary of the transformer for a nominal voltage of 117 V? *Hint:* Add up the voltage drops at the lowest power line voltage. Determine the minimum secondary voltage. Then correct for the line voltage.

ANSWERS

1. 150 mW.

2. The current in the load is 5 mA. The zener current is reduced to 5 mA. The dissipation is 75 mW.

3. The change in voltage is 0.15 V. The change in current is 1.92 mA. The source impedance is 16.3 Ω.

4. The collector current changes 0.5 mA. The base current changes 2 µA. The β is 250.

5. The output changes 1.95 V when the input changes 2 V. The gain is 0.975.

6. The ratio of voltage change to input current change is 2 V/0.1 mA = 20 kΩ.

7. 3.4 Ω.

8. Both loads take 118 mA. The total current is 138 mA.

9. 15 V. The emitter follower is cut off.

10. At the lowest line voltage the secondary voltage must be 15 V + 5 V + 0.6 V + 2 V + 1 V = 23.6 V. The line voltage correction is 117/105 = 1.114. The secondary voltage should be set at 26.3 V.

4 More Semiconductor Circuits

Objectives

In this chapter you will learn:

- about stacked emitter followers
- how an emitter follower can be used to regulate voltage
- about constant current supplies
- how a differential input stage operates
- how a transistor is used as a switch
- about triacs, SCRs, and phototransistors

The Stacked Emitter Follower

Most amplifiers or circuits with gain provide output voltage to various loads. These loads could be a speaker or a long cable. As we saw earlier, the simple npn emitter follower must draw a steady average current in order to supply a negative output voltage to a load. For a negative output voltage, the load current is subtracted from the collector current.

This is a drawback, but it can be avoided by using two emitter followers, as shown in Figures 4.1 and 4.2.

The top transistor is an npn type and the bottom transistor is a pnp type. The top transistor supplies load current when the output voltage is positive, and the bottom transistor supplies load current when the output voltage is negative. The static current through the two transistors when the output voltage is 0 can be limited to a few milliamperes. The symmetry of this circuit allows the output voltage to be 0 when the input voltage is 0.

The operating voltage at the bases of the two transistors can be set so that a small amount of collector current flows when the output is at 0 volts. This is accomplished by a voltage divider using two low-current signal diodes and resistors R_1 and R_2. The forward drop in these diodes

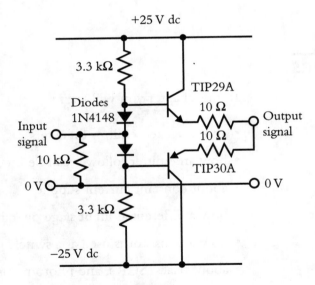

Signal diodes are 1N4148 or equal.

The output npn transistor is TIP29A or equal.
The output pnp transistor is TIP30A or equal.

This circuit can handle a 100-Ω load resistor at a signal of 20 V peak-to-peak. The frequency response is from dc to 100 kHz.

Figure 4.1 A stacked emitter follower

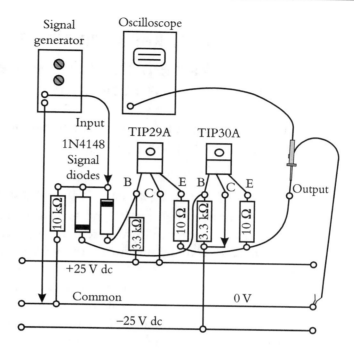

The power supply is not shown on this construction.

Figure 4.2 The construction of the circuit in Figure 4.1

is sufficient to set the base voltages to +0.6 V and −0.6 V. The midpoint of the diodes is the input terminal. This input point is normally connected to a signal from another circuit. In the Learning Circuit I have placed a resistor of 10 kΩ between the input and the circuit common.

The two 10-Ω emitter resistors R_3 and R_4 are required to avoid a possible instability. It is good practice in electronics to avoid paralleling connections between two or more active elements. In this circuit the transistor bases are connected together though diodes. Tying the two emitters together would violate this rule. The 10-Ω resistors in the emitters are a safety factor to remove any possible instability. The instability could be an oscillation above 20 MHz, which could overheat the transistors and yet go undetected.

The input impedance of the emitter followers is high compared to the impedance of the voltage divider. If the β of the output transistors

LEARNING CIRCUIT 25
The Stacked Emitter Follower and How It Functions

You will need (in addition to power supply and measuring equipment):
  1 TIP29A and 1 TIP30A transistor
  2 1N4148 signal diodes
  2 10-Ω, 2 3.3-kΩ, 1 10-kΩ, and 2 100-Ω (not shown) resistors

1. Build the circuit of Figures 4.1 and 4.2 on the circuit board. Check to see that the emitter followers are drawing current. You can measure the current level by measuring the voltage drop across the 10-Ω resistors. If the static current is 10 mA, the voltage is 0.1 V. You can easily measure this using your oscilloscope. This is too small a voltage to measure using the dc scale on a multimeter. If there is no current, then add a resistor of 100 Ω in series with each signal diode (D101 and D102). This will increase the base voltages, slightly allowing the transistors to conduct.
2. Connect a sine wave generator to the input and observe the output voltage. The two signals should be the same.
3. Measure the gain and the output impedance. Try a square wave input signal and note the rise time at 50 kHz. You may want to leave this circuit connected so that you can use it at a later time.

is 150 and the maximum transistor current is 100 mA, the maximum base current is 0.66 mA. The voltage divider current should be about 6 mA. If the divider resistors are 3.3 kΩ, the input impedance is approximately 1,600 Ω. As we will see later, a standard IC amplifier can easily supply current to this load. A signal generator with a peak-to-peak output of 20 V can drive this circuit to show its output capability.

Emitter Followers as Voltage Regulators

An ideal constant voltage source is a voltage that does not vary with a change in load current or a change in power supply voltages. In prac-

> ### ⎍ LEARNING CIRCUIT 26 ⎍
> ### Building a Positive Voltage Regulator
>
> You will need (in addition to a power supply and measuring equipment):
> - 1 TIP29A transistor
> - 1 15-V zener diode
> - 1 100-μF, 35-V capacitor
> - 1-kΩ and 1 330-Ω resistor
>
> 1. Add the positive voltage regulator of Figure 4.3 to your circuit board. The construction is shown in the top half of Figure 4.4. You now have a regulated 15-V source with respect to the common or zero of the circuit. The base voltage is held at 15 V so that the output voltage is also held to near this same voltage. Measure this voltage using the oscilloscope.
> 2. Add a load resistor of 330 Ω from the regulated output (emitter) to the common. Note any change in voltage at the emitter.
> 3. Measure the power supply ripple at the collector of the pass element (25 V). Note the power supply ripple at the output of the regulator. Most of it should be removed. The voltage should not shift more than a tenth of a volt; if it does, check over your connections carefully to be sure they are correct.

tice, voltage regulators are not perfect. In many situations regulation to within 5% is totally acceptable.

You have already seen three examples of practical constant dc voltage sources. The first example was the zener diode circuit of Figure 3.7. The output voltage (the voltage across the zener diode) was fixed. The current for a load was taken from the zener diode itself. In this circuit the zener current must be greater than the maximum load current. If the power supply voltage changes, the zener voltage remains essentially constant.

The second example of a voltage source was the emitter follower in Figure 3.15. Here the emitter output impedance was low. If the base input voltage is fixed, the emitter output voltage is also nearly fixed. If you give this same circuit a fixed input voltage, it becomes a positive voltage regulator, as shown in Figure 4.3.

124 PRACTICAL ELECTRONICS

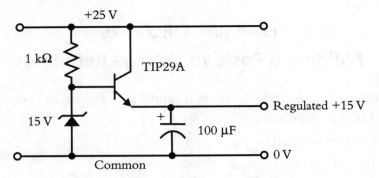

Figure 4.3 A positive voltage regulator

⎯⌁⎯ **LEARNING CIRCUIT 27** ⎯⌁⎯
Building a Negative Voltage Regulator

You will need (in addition to a power supply and measuring equipment):

1 TIP30A transistor

1 15-V zener diode

1 100-μF, 35-V capacitor

1-kΩ and 1 330-Ω (not shown) resistor

Add the circuit of Figure 4.5 to your power supply. The construction is shown in the bottom half of Figure 4.4. Now you have a regulated plus and minus 15-V power supply. This negative voltage regulator should be tested the same way the positive regulator was tested in Learning Circuit 26. Note the voltage at the emitter follower before and after applying a 330-Ω load resistor. Compare the ripple reduction from the collector to the emitter on the pass element. Remember, ripple voltage is the ac component riding in the dc power supply voltage.

The zener current is supplied from the unregulated voltage source through a resistor. This zener voltage defines the base voltage of the transistor. The emitter of the transistor is the output voltage. The transistor functions as an emitter follower, except with a fixed base voltage. When a TIP30A transistor is used, this type of voltage regulator can easily supply 100 mA. The transistor is often called a *pass element,* as the load current must *pass* through the transistor.

The third example of a voltage source was the pnp emitter follower in Figure 3.24. If the base voltage is defined by a zener diode, the output of the emitter follower is regulated. Again the zener current is supplied from the unregulated voltage source through a resistor. The transistor functions as an emitter follower except that the base voltage is fixed. This circuit is shown in Figure 4.5. With a TIP30A transistor, this voltage regulator can easily supply 100 mA. This transistor is also called a pass element.

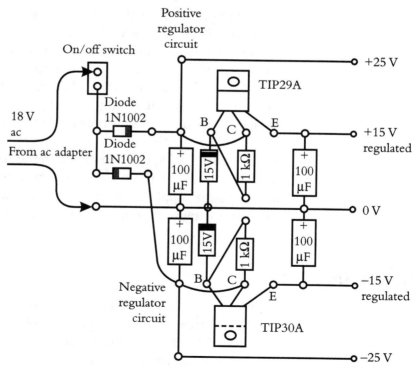

Figure 4.4 The construction of the circuits in Figures 4.3 and 4.5

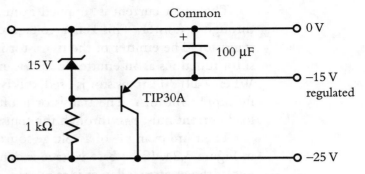

Figure 4.5 A negative voltage regulator

Voltage Sources

The ideal voltage source supplies a voltage that is independent of the load current. The voltage source might be a dc value, or any waveform for that matter. The load could be a resistor, an open circuit, or even a capacitor. If the voltage waveform is a sine wave, then the concept of output impedance can be used. In all practical circuits the regulation varies with frequency. If the source impedance measures 1 Ω from dc to 100 kHz and is 10 Ω at 1 MHz, then the source impedance is inductive. There is no physical inductor. The circuit simply performs as if an inductor were present. In this example the inductance has a reactance of about 10 Ω at 1 MHz.

Electronic voltage sources are different from batteries. It is possible to force current to flow backward into a battery. This is not the case with a power supply circuit involving diodes or transistors. These circuits do not allow current to flow in the reverse direction.

Current Sources

Current sources are not as common as voltage sources. A current source supplies a current that is independent of the load impedance. The load in this case might be a short circuit or a resistor. A constant current source cannot function into an open circuit, as the current cannot flow. The higher the resistance value, the higher the voltage must be for the same current. If the circuit cannot supply the voltage, the circuit cannot function. As an example, assume a constant current of 10 mA. If 10 mA flows in 100 Ω, the voltage is 1 V. If the resistance is 1,000 Ω, the voltage is 10 V. If the resistance is 10,000 Ω, most circuits cannot function.

Surprisingly, we have already encountered a current source. It is the transistor. The collector current is almost independent of the collector voltage. If the collector current is 10 mA, the voltage drop is 1 V for a collector resistor of 100 Ω and 10 V for a collector resistance of 1,000 Ω. The impedance at the collector is the ratio of voltage change to current change. As an example, assume the voltage across a 100-Ω resistor is 1 V and 9.99 V for a 1,000-Ω load resistor. The change in voltage is 8.99 V. For the 100-Ω load the current was 10 mA. For the 1,000-Ω resistor the current was 9.99 mA. The change in current is only 0.01 mA. The ratio of collector voltage change to current change is $8.99/10^{-5} = 899,000$ Ω. This is the collector source impedance. The actual output impedance is the collector impedance in parallel with the collector resistor.

Active Constant Current Sources

Two constant current circuits that are often used in design are shown in Figures 4.6 and 4.7. The first circuit uses an npn transistor to supply a constant current from the positive supply to the common. The second circuit supplies a constant current from the negative supply to the common. In both circuits an emitter resistor and the zener diode voltage determine the current level. If the resistor is 470 Ω, and the zener voltage is 5.1 V, the current is about 10 mA. This current does not depend on the collector voltage, as long as the transistor is within its normal operating range.

A constant current source is not affected by any voltages that are in series with the load. If the resistor is terminated on an ac voltage source at 60 Hz, the current in the resistor is still constant. The 60-Hz voltage cannot add current to a current source. This means that the voltage across the resistor will not have any 60-Hz content.

Voltage sources have a similar quality. A voltage source holds constant for any 60-Hz current that flows in the load. An external source of 60-Hz current cannot change the voltage from a regulated voltage source. Both of these statements are true as long as the circuits are within their normal range of operation.

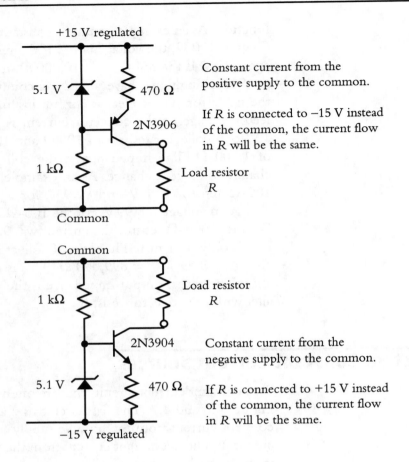

The constant current value in both circuits is controlled by the zener voltage and the emitter resistor. The current in both circuits is about 10 mA. The voltage across a 600-Ω load resistor R is 6 V. If $R = 300\ \Omega$, the voltage will be 3 V.

Figure 4.6 Two constant current circuits

The constant current circuits shown in Figure 4.6 cannot be placed in series. Any slight imbalance in the current levels will upset the circuits. To get around this problem, a load resistor must provide a path for the difference current. There is a similar problem with voltage sources. Voltage sources cannot be placed in parallel without a connecting resistor. This resistor will limit current flow between the voltage sources

More Semiconductor Circuits 129

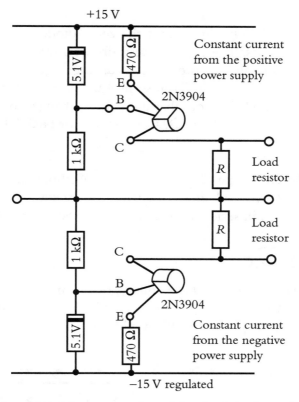

Figure 4.7 The construction of the circuit in Figure 4.6

─⟋\/\⟍─ LEARNING CIRCUIT 28 ─⟋\/\⟍─
Building Two Constant Current Sources

You will need (in addition to a power supply and measuring equipment):

1 2N3904 and 1 2N3906 transistor

2 470-Ω and 2 1-kΩ resistor

2 5.1-V zener diodes

On your power supply board you now have two regulated voltages and a stacked emitter follower. Now you need room for a constant current source. You are going to use the two regulated voltages as the power supply for the constant current sources. Use the circuits in

(Continued)

> Figure 4.6 and 4.7. The current for the zener diodes comes from the 25-V power supplies. The collector resistors are 1,000 Ω. Measure the voltage drop across these resistors. Each current source has been set up to be 10 mA. The voltage drop across the 1,000 Ω is 10 V. Now change the resistor to 470 Ω and note that the voltage changes to 4.7 V. What happens if the resistor is 2,000 Ω? Can the resistor be 3,000 Ω?

based on the difference voltage. Without this resistor, voltage sources will overload or self-destruct. To review, current sources cannot be placed in series and voltage sources cannot be placed in parallel.

The Differential Stage

The differential stage consists of two matched transistors. A typical circuit is shown in Figures 4.8 and 4.9. This circuit is usually used as the input stage in most IC amplifiers.

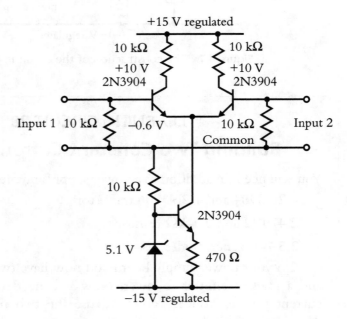

Figure 4.8 A differential stage and a constant current source

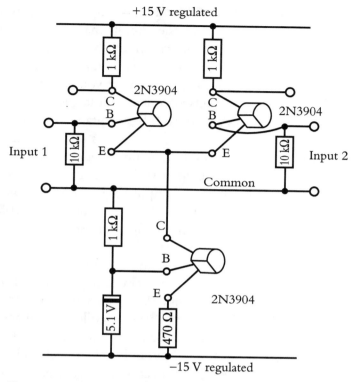

Figure 4.9 The construction of the circuit in Figure 4.8

We will call the transistor base on the left the input or first base. The two emitters are connected to a constant current source, Q3. In this example the two base voltages are connected to the zero reference potential through 10-kΩ resistors. If the transistors are well matched, then half the available constant current flows in each transistor. The voltage across each collector resistor is equal to half the constant current times the resistance value. The collector voltage is equal to the supply voltage minus this voltage drop. In Figure 4.8 the total current is 1.0 mA. The collector resistors are 10 kΩs. The voltage drop across each resistor is 5 V. If the supply voltage is 15 V, the collector voltages are each 10 V.

The input base current is equal to the collector current divided by the beta of the transistor. If the beta is 250, then the base current is 0.5mA/250 = 2 µA. This current must be supplied or the transistors cannot function. In this example the input base could be supplied base current through a resistor connected to the power supply. This resistor is $R = 15/2 \times 10^{-6} = 7.5$ MΩ. In most applications this base current is supplied by the circuits that connect to the bases.

The operation of this differential stage can be demonstrated by applying different voltages to the two input bases. If the two bases are set to +3 V, the emitter voltages follow. The current is held constant so that the two collector voltages remain at 10 V. In other words, there is no gain. If both base voltages are set to −3 V, the emitters again follow and the collector voltages remain unchanged.

If the first or input base voltage is raised to 0.01 V and the second base is held at 0 V, the emitter voltage splits the different and rises to 0.005 V. The input base emitter voltage is now 0.005 V and the second base emitter voltage is reduced by 0.005 V. The result is that more current flows in the first transistor and less current flows in the second transistor. Because of the constant current source, the sum of the two collector currents is a constant. The change to the collector voltages depends on the gain of the transistors. In this example, assume the collectors change 0.5 V. The first collector voltage will drop from 10 V to 9.5 V and the second collector voltage will rise from 10 V to 10.5 V. With this information we can calculate the voltage gain. The input changed 0.01 V and each collector changed 0.5 V. This is a gain of −50 to the first collector and a gain of +50 to the second collector. The gain is associated with an offset of about 10 V.

A matched pair of transistors can be purchased as a single component. This matching can be significantly better than the matching provided by the transistor pair at the input to a standard IC amplifier. The careful matching is necessary in high gain amplifiers where dc drift must be very low. Some manufacturers of IC amplifiers balance the input stages as a part of the manufacturing cycle.

The differential pair has no gain when both bases were raised or lowered in potential by the same amount. A signal that is common to both inputs is called a *common-mode signal*. In this circuit the common-mode gain is practically 0. In effect, the common-mode signal is rejected. The gain to the difference signal is called *normal-mode gain*.

To illustrate how a differential stage separates normal-mode signals from common-mode signals, consider the gain to a signal where the input to base one is 3.01 V and the input to base two is 3.00 V. The average input signal is 3.005 V. The gain to this common-mode signal is 0. The input difference signal or differential signal is 0.01 V. The gain to this difference signal is −50 to the first collector and +50 to the second transistor.

⌁ LEARNING CIRCUIT 29 ⌁
Building a Differential Input Circuit

You will need (in addition to your circuit board and measuring equipment):

3 2N3904 transistors

1 5.1-V zener diode

4 10-kΩ, 1 1-kΩ, and 1 470-Ω resistor

Build the circuit shown in Figures 4.8 and 4.9. Note that it does not require a matched pair of transistors to illustrate the principle. The regulated ±15-V power supplies are used for this circuit. The constant current source is set for 10 mA. After you get the circuit built, make sure that the two collector voltages are about 10 V each. Since the gain can be 50, any input signal from a generator should be attenuated by about 100:1 so that the signal level can be easily adjusted. A suggested voltage divider for each base is a series 100-kΩ resistor and 1-kΩ resistor. This divider attenuates the signal from your function generator. When a 1-V sine wave signal generator is applied between each 100-kΩ resistor and common, the signal on that base is about 0.01 V. After gain, the collector signal voltages should be about 1 V. The same signal level should appear on each collector.

If your oscilloscope has A and B inputs, you can sum the two collector signals. Be sure the oscilloscope inputs are ac coupled. If the result is 0, it proves the signals are of opposite polarity or balanced. Tie the two 100-kΩ resistors together and connect the signal generator between these two resistors and common. Observe the collectors for signal. This is the common-mode test. The collector signal should be very small.

In our example of gain, the input to the first base is a signal with respect to the zero reference conductor. This is called a *single-ended signal*. The two collector signals are equal and opposite in polarity. Remember, we must ignore the static operating voltage of 10 V. The input signal changes from 0 to +0.01 V and the output changes ±0.5 V around the nominal operating voltage. The two output signals are called a *balanced signal*. A balanced signal pair means that one signal goes positive when the other goes negative. For a balanced signal pair the average value is 0. Balanced signals are often used in driving long cables. Certain kinds of noise coupling can be eliminated by using this technique.

Field Effect Transistors

So far we have considered junction transistors. Another form of transistor is called the *field effect transistor, or FET*. This type of transistor is also formed from p and n silicon material. The current path in a FET transistor is called a *channel*. The channel can be all p or n material. There are no junctions through which this current must flow. The channel current flows directly from the source to the drain in the channel. The ends of the channel are called the *source* and the *drain*. These designators correspond directly to the emitter and collector in a junction transistor.

The controlling element is called a *gate*. The voltage between the gate and the source controls the current flow through the channel. The gate is actually insulated from the source and the drain. When the gate is formed as an insulated junction, the device is called a *junction FET*, or a *JFET*. When the gate is insulated by a metal oxide, the device is called a *MOSFET* or *m*etal *o*xide *s*emiconductor *f*ield *e*ffect *t*ransistor. There is gate current, but it is usually nanoamperes instead of microamperes. For an n-channel FET the drain voltage is positive with respect to the source.

There are four basic types of FET devices. The conducting channel can be n or p material and the devices can be enhancement or depletion types. The enhancement FETs do not conduct unless there is a gate voltage. The depletion mode devices allow current to flow at 0 gate voltage. Some devices have both enhancement and depletion regions of operation. For all FETs, the gate voltage controls the electric field along the channel. It is this field configuration that controls the flow of current in the channel. The absence of a pn or np junction means that the voltage drop from the source to the drain can be very small. When the transistor is turned fully on, the channel is a low resistance. When the channel is turned off, it is a very high resistance. This makes the FET an ideal switching device.

A FET does not have beta, as there is no controlling gate current. The voltage curves for a typical n-channel JFET device are shown in Figure 4.9. In most circuits the FET can be treated very much as a junction transistor. There are FET followers, FET gain circuits, and FET pass elements in power supplies. There are IC amplifiers made from FET devices. If you understand how transistors are applied to a circuit, the FET should offer you no difficulty. For this reason we limit our Learning Circuits to transistors. A problem can arise in using depletion mode devices. The needed gate voltage may extend beyond the available power supply voltage. Typical operating curves for a FET transistor are shown in Figure 4.10.

There are logic circuits made from FETs. In one family of devices both p and n channel material are used in the form of stacked emitter followers. These components are known as CMOS, which stands for *c*omplementary *m*etal *o*xide *s*emiconductor.

The symbols used for FETs are sometimes confusing. Manufacturers attempt to show the construction of the device in the symbol they use for it, but these are not always quite clear. A few of the common symbols are shown in Figure 4.11.

FET devices are particularly sensitive to handling; any slight electrostatic discharge (ESD) can destroy the component. For this reason, FET devices are shipped in conductive plastic to limit generating a charge when the components rub on the plastic. An accumulated charge

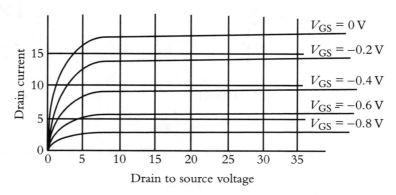

V_{GS} = gate to source voltage

Figure 4.10 The operating curves for an n-channel JFET transistor

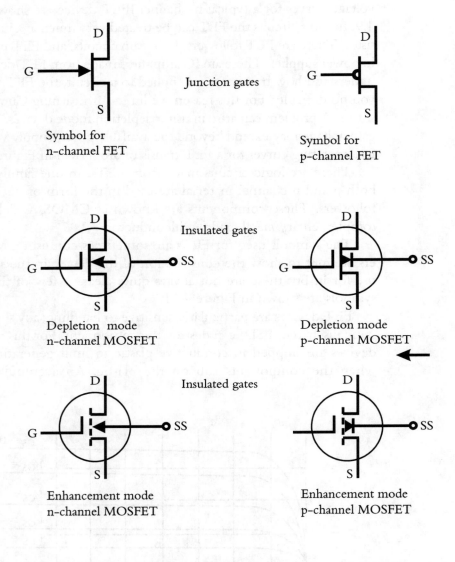

SS is an abbreviation for substrate. The substrate is the material under the deposition of the transistor elements.
G = gate, S = source, D = drain

Figure 4.11 Symbols used for FET devices

represents an electric field that can cause arcing. The leads of a FET device should not be touched when handling. It is bad practice to throw FET devices into a pile with other components and then attempt to sort them out by hand.

Light Emitting Diodes

When a pn junction is properly doped, the current in the junction can generate light. Light results when electrons release energy upon returning to their normal state from a higher energy state. The color of the light is directly related to the energy levels associated with the two states. The light emitting diode or LED operates in the forward direction of the diode. LEDs are not designed to accommodate a high reverse voltage. Typical LEDs operate in the current range 5 to 20 mA. Figure 4.12 shows how a transistor can be used to turn on an LED. The collector resistor is necessary to limit the LED current.

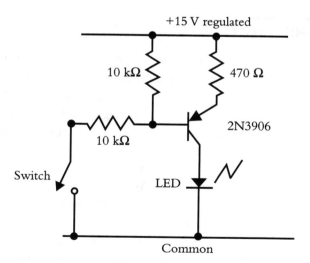

When the switch closes, the base voltage is about 7.5 V. The emitter follows to about 8.1 V. This is a transistor current of 15 mA. This current flows in the LED providing light.

Figure 4.12 A transistor used to turn on an LED

138 PRACTICAL ELECTRONICS

⊸⟋⟍⊸ LEARNING CIRCUIT 30 ⊸⟋⟍⊸
Building a Transistor Switch

You will need (in addition to a power supply and measuring equipment):
1 SPDT switch (used as a SPST switch)
1 2N3906 transistor
1 LED diode
2 10-kΩ resistors and 1 470-Ω resistor

Construct the circuit shown in Figure 4.12. When resistor R_1 is connected to the plus supply, the LED is turned on. I have not provided a construction diagram, as you are now experienced enough to lay out this circuit yourself. Congratulations on reaching this milestone, which indicates how much you have learned from carefully constructing the circuits.

Phototransistors

When light strikes a semiconductor material, the absorbed light energy (photons) can knock electrons out of their normal positions. In the base region of a transistor, these free electrons allow transistor action. If there is a collector voltage, then collector current results. This is the reason why semiconductor devices are constructed in plastic in such a way as to limit any light entry.

A phototransistor is a device that intentionally uses light to control a transistor. An LED is positioned to shine light into the base region of a transistor. The light from the LED generates free electrons. The transistor reacts to these free electrons as if there were base current.

The LED is electrically separated from the transistor. In fact, the LED and the transistor can be in completely separate circuits. It is possible for the transistor to be associated with an ac power conductor and the LED to be associated with a circuit on the secondary of a transformer. This separation in function makes a phototransistor valuable. The voltage between LED and the transistor can be as great as 500 V. Because of the separation, the LED and the transistor can be drawn in separate parts of the circuit diagram. See Figure 4.13.

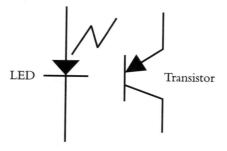

The symbol for a phototransistor

The LED shines directly into the base region of the transistor. This light provides free electrons in the transistor that are the equivalent of base current. The LED diode and transistor are insulated from each other and can be used in different circuits.

Components are available that provide several transistor/diode pairs. These components may connect all the emitters together or all the LED cathodes together to limit the pin count.

Figure 4.13 A phototransistor symbol

The Signal Controlled Rectifier or SCR

A diode always conducts in the forward direction. A signal controlled rectifier, or SCR, functions like a diode with a control gate. SCRs find most of their applications in ac power circuits. The gate can control the SCR so that current can start flowing over any portion of the conducting half-cycle. Once current starts to flow, the gate loses control until the current returns to 0.

The SCR is constructed of four layers of doped silicon. The layers are pnpn. This arrangement of semiconducting material is known as a Shockley diode. The outer two layers are the p and n material of a standard diode. The outer n material is the cathode. The center np material is the SCR gate. The forward-conducting direction of the diode is to the outer n layer. When the gate is made positive with respect to the cathode, the SCR will conduct in the forward direction. If the gate is at the cathode potential, the SCR will not conduct. The symbol and the construction of an SCR are shown in Figure 4.14.

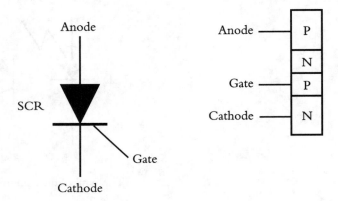

The diode conducts in the forward direction when the gate is positive with respect to the cathode. The firing voltage can vary from 5 V to 30 V.

Once the SCR is turned on, the current must return to 0 before the SCR can be turned off by the gate.

Figure 4.14 The symbol and structure of an SCR

The Triac

A triac consists of two SCRs in parallel. The first SCR controls the first half–power cycle and the second SCR controls the second half–power cycle. The gate voltage must be positive with respect to the corresponding cathode to turn the triac on in that half-cycle. The symbol and construction of a triac are shown in Figure 4.15. The word "triac" is a coined word implying three states. These three states are *off* and the two directions of *on*.

Triacs can be controlled by a phototransistor connected between the gate and one of the triac terminals. When the LED emits light, the triac turns on. The response time of the triac is in the order of microseconds.

Triacs are used in most household light dimmers. An RC delay circuit is used to turn the triac on during each half-cycle. Changing the resistor value controls the point in the cycle when the triac conducts. A simple version of this approach is shown in Figure 4.16.

When the voltage between the gate and terminal 1 exceeds a threshold voltage, the triac conducts. When the adjustable resistor (poten-

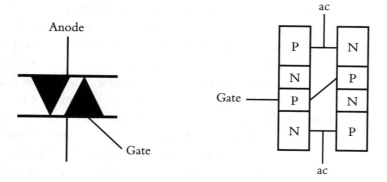

A triac is made of two SCRs back to back. The one gate controls both SCRs.

The term *triac* is a coined word meaning three ac states. The states are off and both directions of on.

Figure 4.15 The symbol for a triac and its construction

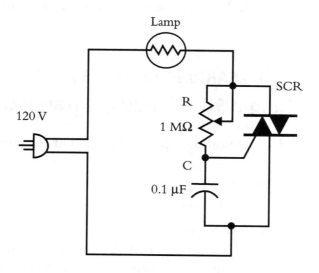

Warning: Do not attempt to build this circuit. There is a danger of shock.

Figure 4.16 A light dimmer circuit using a triac

tiometer) is set to 0 Ω, the triac conducts for the entire cycle. As the resistance increases, the capacitor delays voltage on the gate and the triac conducts later in each half-cycle. The resistor and capacitor values must be set so that at maximum resistance the triac never turns on. Note that during the time the triac is conducting, the line voltage appears across the lamp and the capacitor is discharged. There are many versions to this circuit, including a time delay using two resistors and capacitors, bias voltages applied to the gate, and resistors or zeners in series with the gate. These circuits are used to accommodate different triac characteristics.

When a triac is turned on in midcycle, a step voltage is applied to the load. This type of signal can cause interference in nearby electronic devices. For this reason, line filters are often a part of a light dimmer design. The filter is often a small series inductor and a shunt capacitor that slows the rise in load current.

Transistors as Switches

We have discussed the transistor as a component that can provide gain. When the transistor is used at the extremes of its operation, it can serve as a switch. When there is an excess of base current, the collector cur-

---

LEARNING CIRCUIT 31
Building a Switch Using a Transistor

You will need (in addition to a power supply and measuring equipment):

1 2N3904 transistor

1 1N4148 and 1 LED diode

1 1-kΩ and 1 10-kΩ resistor

Build the circuit shown in Figure 4.17 on your board, using your own layout. Connect a square wave generator set to 3 Hz to the control resistor. If the square wave is symmetric about 0 V, then clamp the input base so that it cannot go negative. The arrow of the diode goes to common. The LED will turn on three times per second.

rent increases until there is a small collector voltage. When the base current is 0, the collector current is 0 and the collector voltage is at a maximum. This is the same as using a mechanical switch. When the switch is open there is no voltage across the resistor, and when the switch is closed the power supply voltage appears across the resistor.

A switch contact might easily handle several hundred milliamperes. In our example this current level in a transistor might be damaging. An open switch contact can easily handle 100 V. In a transistor this might exceed the collector breakdown rating. In these two ways the mechanical switch is superior in performance. But the transistor is faster and less expensive, and it also has a longer life. A transistor switch can operate in a microsecond, while a relay contact might take 5 ms. This is 5,000 times slower. The transistor is small, the amount of operating power required is very low, and there are many switching tasks that are within its rating. It is this switching ability that makes computers possible. This will be discussed in chapter 6.

An example of a transistor switch is shown in Figure 4.17. The LED is turned on when the input voltage is positive. The LED can be made to toggle on and off by using a square wave voltage generator set to a frequency of 2 or 3 Hz. This is a simple circuit that you can build and operate.

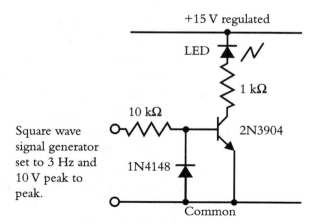

The diode clamps the square wave voltage so that the base is not driven negative.

Figure 4.17 An LED that is toggled on and off using a transistor switch

The Switching Power Supply

Transistor switches can be used to change a dc voltage into an ac voltage. The ac voltage can be associated with a step-up or step-down transformer. If the transformer has several secondary coils, then several new ac voltages can be obtained. These secondary voltages can be rectified and used to supply dc power to separate circuits. The circuit in Figure 4.18 uses two transistor switches and one secondary voltage. Later we will discuss how the signals are generated that drive the transistor switches.

When transistor Q1 is turned on, transformer terminal 1 is effectively connected to the power supply common lead, or 0 V. The centertap 2 of the transformer primary is connected to the positive supply. This places the power supply voltage across one half of the primary coil. The B field in the core material of the transformer starts to increase. In one half-cycle transistor Q1 is turned off and transistor Q2 is made conductive. This results in placing the power supply voltage between the centertap 3 and terminal 2 of the transformer. The B field in the core now reverses direction. At the end of the cycle transistor Q2 is turned off and transistor Q1 is again turned on. This constant toggling of the two switches produces a square wave voltage across the transformer primary coil.

When the voltage on terminal 1 of the transformer is at 0 V, terminal 2 is at twice the power supply voltage. Current does not flow in this half of the primary because transistor Q2 is turned off. On the next half-cycle the same thing happens to transistor Q1. Terminal 2 is at double the power supply voltage when terminal 2 is at 0 V. The transistor must be rated to withstand double the power supply voltage.

The square wave voltage on the secondary coil of the transformer can be rectified to form a new dc power supply. If the ratio of secondary turns to primary turns is 2:1, the new voltage will be double the original power supply voltage.

This circuit is at the heart of many switching power supplies used in today's electronics. The switching frequency can be 50 kHz, making the transformer a small component. At this frequency the power supply filter capacitors can also be fairly small.

There are several additional points regarding the circuit in Figure 4.18. On the first cycle the core material starts out with the induction flux $B = 0$. The means that the core material may saturate on the first

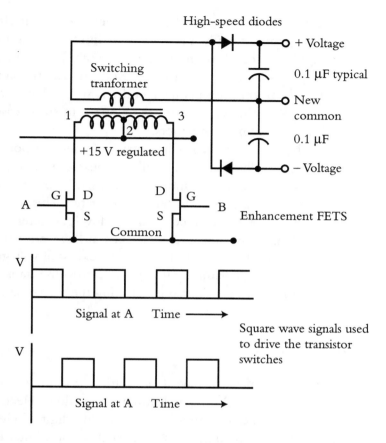

When signal A is positive, transistor A is turned on, 15 V are impressed between transformer terminals 1 and 2, and transistor B is turned off. The voltage at point 3 rises to 30 V. In the next half-cycle transistor B is turned on, transistor A is turned off, 15 V impressed between terminals 3 and 2 of the transformer, and terminal 1 rises to 30 V. The square wave frequency might be 20 kHz. If the number of turns on the transformer primary is 30, there is 1 V per turn. If the secondary has 20 turns, the secondary voltage is 20 V. The dc output voltages from the secondary are plus and minus 20 V less the diode drop.

Figure 4.18 Two transistor switches used to convert dc to ac and back to dc in a second circuit

cycle. Within a few cycles the B field centers itself so that the field moves symmetrically between two limits. The transformer must be designed so that the core material, the number of turns, and the applied voltage result in a B field that never saturates the core material. Typically, at 50 kHz the primary coil has somewhere between 10 and 20 turns. With this small number of turns, in some cases it is possible to wind your own transformers.

The transistors will self-destruct if they are both turned on at the same time. The transistors will overheat even if this time period is as short as 1 µs. One technique that is used is to slow down FET transistors by using a resistor in series with gates. This makes use of the FET transistor gate capacitance to form a low-pass filter.

When a transistor interrupts current flow to a transformer, there can be a significant voltage overshoot. This excess of voltage can often be limited by placing a damping RC circuit across the transformer. This excess voltage must be considered in selecting the transistor voltage rating.

Digital Switching

The electronics we have been studying, and will continue to study throughout this book, is known as "analog" electronics. Electronics also has a quite different branch, called "digital" electronics, which is not part of this book. However, analog and digital electronics are not completely separate in actuality. Every digital device uses analog electrical signals, and there are countless areas where the two branches of electronics must work together. It is often necessary to convert analog signals to a digital format, and vice versa. For this reason, I have included this brief section on digital switching, even though it is not our main topic. Transistors used as switches often provide an interface to digital devices. There are special IC components, made up of large numbers of transistors, that perform these tasks. We will look at them in chapter 6.

"Analog" is defined (in the IEEE *Standard Dictionary of Electrical and Electronics Terms*) as "describing a continuously variable physical quantity, such as voltage . . . that normally varies in a continuous manner." "Digital," by contrast, proceeds through *logic,* in which change is considered as occurring in a stepwise way, rather than in a continuously variable curve. The difference will be clear if you think of two kinds of clocks, one with hands moving around a face, and one with numbers. The clock with hands moving continuously is an analog time-measuring device. Time moves continuously, and so does an analog clock. The clock with numbers is a digital clock. This clock tells you the time is 2:46, until the moment when it suddenly tells you it is 2:47. There is nothing in between; it's one time or the other. This is logic. Is it 2:46? One logic state, called logic 1, says yes. The other logic state, logic 0, says no. If it is not 2:46, then it is some other time completely.

In digital circuits, the signal is either on (logic 1) or off (logic 0.) There is no current or voltage representing any intermediate state. There is no resistance, capacitance, or inductance, or any of the analog concepts we have been studying so far. However, any digital device still needs analog power to make it run. The power supply running that digital clock is still basically the same kind of power supply you have already studied—an analog device.

Electronic designers must provide many interfaces between analog and digital signals. Digital designers tend to focus exclusively on their digital world, and leave the problem of interface to analog designers. So, while we are studying transistors as switches, it makes sense to take a look at the way transistors used as switches can be used in these interface situations.

Digital signals are usually power supply voltages. A logic 1 might be the positive 5 V and a logic 0, 0 V. The exact voltage is not important; a voltage of +4.5 V would still be considered a logic 1 and a voltage of +0.5 V a logic zero. A logic signal can be developed by turning a transistor on or off. As an example, assume a transistor clamps a logic conductor + 5 V for a logic 1. When the clamp is turned off, a resistor connects the conductor to 0 V, or logic 0. This is known as a *pull-down resistor.*

The opposite is also possible. A transistor clamps a logic conductor to 0 V. When the transistor is turned off, a resistor connects the logic to +5 V. This is known as a *pull-up resistor.* The clamping action is what generates a logic signal. If the conductor is not clamped in the first circuit, it can be clamped by a second circuit. This allows the same logic conductor to receive logic rather than send logic. A control line can be used to define the direction of transmission. Obviously, a logic conductor cannot be clamped to two different voltages at the same time. The rules at the interface must be understood before any connection is attempted.

Within a logic circuit, a conductor carrying a logic signal may be connected to several devices. Each connection may require current flow, depending on the logic state. This is called *fan-out*. When a logic circuit must draw current from a second device, the logic must be able to "sink" (accept current flow) current. This is an important consideration when transistors are used as switches to interface a logic circuit. Since there are many different logic families, there are many interface problems.

Digital logic is usually controlled by a clock signal. Devices change state when the clock voltage rises or falls. Clock voltages must rise and fall within a given time period, or logical malfunctions can occur. For example, a clock voltage that rises in 1 μs may be too slow to operate some types of logic. If for any reason you must supply a clock signal to a logic circuit, the rise and fall time specification is an important consideration.

SELF-TEST

1. A constant current source provides 5 V into 1,000 Ω and 9.992 V into 2,000 Ω. What is the source impedance?

2. A constant voltage source provides 5 V into 1,000 Ω and 4.992 V into 500 Ω. What is the source impedance?

3. A 50-kHz square wave voltage at 10 V peak is half-wave rectified. How large must the filter capacitor be if the voltage can sag 0.1 V in a half-cycle for a 200-mA load?

4. A differential stage has a voltage gain of 40 from each emitter to each collector. The collector voltages for equal input signals are 9 V. The emitter to base voltage is always 0.6 V. Assume the positive input is 2.5 V and the negative input is 2.45 V. What is the emitter voltage without signal? What is the emitter voltage with signal?

5. What are the two collector voltages in problem 4?

6. In problem 4, reverse the input signals. What are the two collector voltages?

7. Solve problem 5 assuming the positive input terminal is 2.55 V and the negative input terminal is 2.50 V.

8. In problem 4, change the constant current source so that the collectors are 8 V. What changes? What are the collector voltages when the signal voltages are applied?

9. A constant current source supplies a current of 1 mA to a capacitor of 0.1 µF. How long does it take for the voltage to change from 3 to 5 V?

10. A constant current source provides a sinusoidal current of 1 mA at 10 kHz. What is the voltage across a 1-kΩ resistor? What is the voltage across a 0.01-µF signal at 20 kHz?

ANSWERS

1. The voltage change is 5 V. The current change is 0.004 mA. The impedance is 1.25 MΩ.

2. The change in voltage is 0.008 V. The change in current is 5 mA. The impedance is 1.6 Ω.

3. The charge flowing in 10 µs is $I \times t = 0.2A \times 10^{-5}s = 2$ µC. The capacitance equals $q/V = 2 \times 10^{-6}/0.1 = 20$ µF.

4. The emitter voltage without signal is –0.6 V. With 2.5 V on each input, the emitters are at 1.9 V. With 2.5 V on one base and 2.45 V, on the second base the input signal to be amplified is 0.05 V. The emitters move half the distance, to 2.475 – 0.6 = 1.875 V.

5. The difference signal at the input is 0.05 V. The signal change on the first base is +0.025 V and on the second base, –0.025 V. The first transistor multiplies this input signal change by –40. This means the first collector changes from 9 V to 8 V. The second collector changes from 9 V to 10 V. The collector signals are 1 V. The difference signal between the collectors is 2 V.

6. Reverse the input signal and the output signal then reverses. The two collector voltages are 10 and 8 V respectively.

7. The answers remain unchanged. The new signal has added common-mode content, which is rejected.

8. The new collector voltage implies that the constant current source was increased. The gains remain unchanged. The collector signals are still 1.0 V.

9. The charge difference is $q = CV = 10^{-7} \times 2 = 0.2$ microcoulombs. $I \times t = q$. $t = q/I = 0.2 \times 10^{-6}/10^{-3} = 0.2$ ms.

10. The voltage across the resistor is 1 V. The reactance of the capacitor at 20 kHz is 796 Ω. A constant current of 1 mA (rms) by Ohm's law implies a voltage of 7.96 V (rms).

5 Feedback and IC Amplifiers

Objectives

In this chapter you will learn:

- about integrated circuit (IC) amplifiers
- the concept of feedback, and how it is used to obtain positive and negative gain
- how a differential amplifier works
- the way active filters are constructed using integrated circuits

In chapter 4 we completed our survey of the basic circuits that make up a linear integrated circuit (IC) amplifier. These included the input circuit (generally a differential transistor pair); the output stage (generally a stacked emitter follower); and cascading pnp and npn transistors, which provide gain. This configuration of components is useful in a wide range of applications. So you might think that our next step would be to construct this useful amplifier. However, we will not be doing this, for the simple reason that IC manufacturers have designed literally thousands of circuits of this kind. They are small, inexpensive, and easy to obtain, so there is no need for you to construct one yourself. We studied the inner workings of the IC amplifier only so that you would understand what is going on inside them, something even experienced electronics designers do not always know.

IC manufacturers have designed circuits with variations to satisfy a wide range of applications. Some of the designs are intended for high-frequency applications such as video amplifiers used in TVs. Some of them have very low dc drift, so these can be used for amplifying very small signals—for example, in a postal scale. Some can operate using very low operating voltages; others can operate using high-voltage power supplies. ICs are so common and so inexpensive that designers rarely use a single transistor to obtain gain. You would have a difficult time improving on the designs that are available.

Most linear integrated circuit amplifiers are general-purpose components. They are not intended for applications where you connect an input signal to one terminal pair and expect to get an output on another terminal pair. They need to be a part of some additional circuitry in order to be useful. The resulting circuit can satisfy a wide range of applications.

A typical IC has a gain from the input to the output that is often over a million at dc. An input attenuator of a million to one would not be a good way to use this huge amount of gain. It must be dealt with by imbedding the IC in a circuit that regulates the gain. The method used for this purpose is known as *feedback*.

Manufacturers of IC amplifiers and designers who use them must be aware of this problem of regulating gain. The manufacturer must design the IC amplifier in such a way that feedback circuits can be used with it, and the user must understand how to use feedback. Feedback provides so many benefits that it is really worth learning how it works and how it is used.

Negative Feedback

We all use feedback systems automatically whenever we have something we need to regulate or hold steady. For example, when driving a car, we use feedback to hold the speed steady regardless of wind, grade, or road condition. If the speed rises, we reduce the pressure on the gas pedal. Going uphill, when the car slows we add pressure to the gas pedal. This is known as *negative feedback control,* because the input is in the opposite direction to the way the car's speed is changing. If the car is slowing, you make it go faster, and vice versa.

The basic requirements for a feedback control system are an objective, a means to control the objective, and a way to measure the departure from the objective. With speed control, your objective might be to drive the car at a steady 55 mph. The means of control is the gas pedal, which controls the engine. The departure from the desired speed is measured by the speedometer. In an electronic feedback circuit these same three requirements must be present. The objective might be to have an output from the circuit of 10 V. The means of measuring the difference between the desired and the actual voltage (the error voltage) is the amplifier, and the means of controlling the voltage is the gain of the amplifier, which you use to increase or decrease the voltage as needed. If there is enough signal gain, the error voltage can be kept very small.

The Language of Feedback

A feedback circuit is a way to use amplifier gain to meet a desired objective. There is one more feature to a feedback circuit that is not part of ordinary feedback systems such as our example of the car. It is this: A feedback circuit provides a way to subtract or compare a fraction of the output signal with the input signal. Analyzing the difference between a fraction of the output and the input allows the feedback circuit to control gain by using resistors. The difference signal is amplified by the gain of the amplifier to provide the desired output signal. This technique enables you to use the high gain from the amplifier to achieve accurate control of signal gain, as you will see from the following discussion.

To be effective, feedback circuits should have a great deal of excess signal gain. This gain is called *forward gain*. Another expression that is commonly used is *open-loop gain*. This is the signal gain if the feedback is removed. The gain with the feedback present is called the *closed-loop gain*. The *open-loop gain* that is a factor greater than the *closed-loop gain* is called the *feedback factor*. As an example, if the *open-loop gain* is 1,000,000 and the *closed loop gain* is 100, the *feedback factor* is 10,000.

The IC Package

Before we can proceed with our study of IC amplifiers, you need to become familiar with the notation used in schematics for circuits that include them. The symbol for a linear integrated circuit amplifier is a triangle, as shown in Figure 5.1.

The two input terminals are labeled + and −. The power supply voltage connections are usually not shown on this symbol. The output terminal is usually taken from the right at the tip of the triangle. The internal components that make up the IC are connected between the plus and minus power supply voltages. The common power supply conductor is not connected to the IC amplifier. This conductor is used in the external feedback circuit that supports the IC amplifier.

LEARNING CIRCUIT 32
Mounting an IC Amplifier DIP Socket

You will need (in addition to your circuit board):
1 8-pin DIP plastic socket for an IC

There are many package configurations for an IC amplifier. The package we will use in our lessons is called a plastic DIP with 8 leads. The acronym DIP stands for *dual in-line package.* This IC component has two complete independent IC amplifiers. It is normally inserted or mounted into plated through holes in a circuit board and soldered. For this Learning Circuit, simply plug the component into a standard DIP socket. See the diagram for mounting the DIP on Figure 5.3. This allows you to solder to the leads of the socket, so you can replace the IC amplifier without rewiring or soldering to the actual terminals of the IC. Also, with a socket you are less apt to overheat the IC. Use the regulated 15-V power supply voltages. Check to see that the correct voltages appear on the pins of the socket. Do not plug in the IC amplifier at this point.

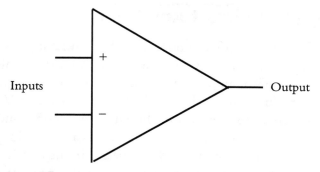

Symbol for an integrated circuit amplifier

The LF353 has two independent amplifiers.

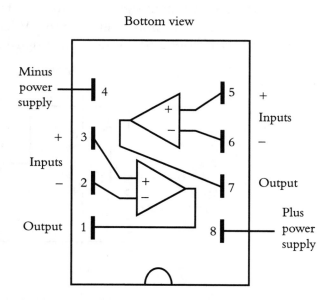

The pin arrangement of a dual IC amplifier. Notice the key at the bottom of the package.

Figure 5.1 The symbol for an integrated circuit amplifier, and the packaging of a dual amplifier

The Gain 1 Integrated Circuit Amplifier

The IC amplifier consists of a differential input stage, some intermediate gain, and an output stage. The input terminals are marked + and −. The power supply connections to the IC amplifier are plus and minus. For our example these voltages are set to ±15 V dc. We start with the simplest possible circuit, shown in Figures 5.2 and 5.3.

Assume that the IC amplifier has an open-loop gain of 1,000,000 at dc. In this application the negative input terminal is connected to the output terminal. Next we connect the positive input terminal to the power supply common through a resistor of 10 kΩ. We do this because the input terminals to any amplifier must not be left open and undefined. This circuit arrangement is a positive "gain 1" amplifier. The gain of 1 is measured from the positive input terminal to the output terminal. If the positive input signal is set to +2 V, the negative input terminal and the output terminal must rise to 2 V less 2 μV. The circuit is in balance because the 2 μV of difference signal is the right signal to be

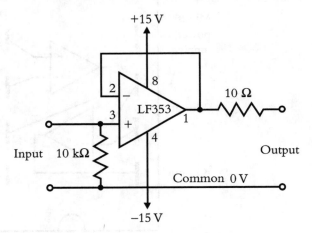

The power connections are not always shown in a schematic. The 10-Ω resistor provides stability protection when there are capacitive loads. The 10-kΩ resistor provides a path for base current if it is not provided by the signal source. Notice that the common lead does not have a direct connection to the In the LF353, either IC amplifier can provide this same circuit.

Figure 5.2 A gain 1 feedback amplifier

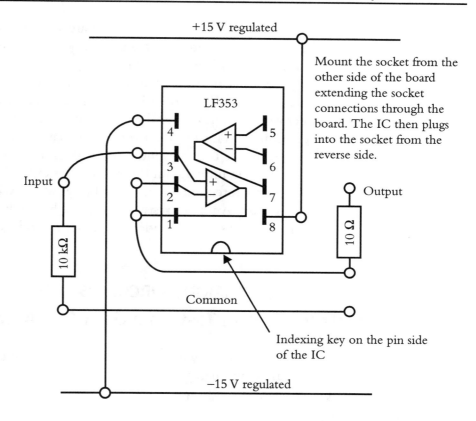

Figure 5.3 The construction of the circuit in Figure 5.2

multiplied by 1,000,000 for an output of 2 V. To be accurate, the gain is actually 0.999999. If the input is set to −2 V, the output terminal will adjust to −2.0 V plus 2 μV, or −1.999998 V.

The circuit in Figure 5.2 has 100% negative feedback. The positive input terminal to the IC is the input terminal to our circuit. This configuration is called *potentiometric feedback*. To see how this circuit operates dynamically, place a +2-V signal on the positive input terminal. The output starts to rise. As the output nears 2 V, the negative input nears the value of the output. Remember, this amplifier reacts only to differences. If the output were exactly 2 V, the input signal difference would be 0. This implies that the output is also 0. In fact, the output can only rise to a final value of 1.999998 V. Obviously, this is very close to the right answer.

The output voltage of this gain 1 amplifier will follow the input voltage as long as the voltage is within the performance specifications of the amplifier. If the voltage is a square wave or sine wave, the output will follow. The circuit acts like a very accurate emitter follower. The output of this IC amplifier will supply up to ±10 mA and ±10 V.

The input impedance to this circuit is very high. When the input base changes 1 V in potential, the emitter also changes 1 V to within a microvolt. The input signal current hardly changes when the input voltage changes. This is a high-input impedance circuit by our definition. There must be base current, but this does not count as signal current.

In the circuit diagram there is a series output resistor of 10 Ω. This resistor is protection against instability. More will be said about this resistor later. An input resistor connects the input terminal to the common. This resistor is provided so that when you are not using the circuit

LEARNING CIRCUIT 33
Constructing and Testing a Gain 1 IC Amplifier

You will need (in addition to your circuit board with mounted socket and your measuring equipment):

1 10-Ω and 1 10-kΩ resistor

1 LF353 IC amplifier

1. Add the circuit shown in Figures 5.2 and 5.3 to your circuit board.
2. Plug the IC amplifier into the socket.
3. Turn on the power. Note the output voltage of the IC amplifier with your oscilloscope. The output voltage should be very close to 0 V.
4. Connect a function generator between the positive input terminal and the common. Always make sure the common of the function generator connects to the common of the power supply.
5. Set the input voltage to 2 V at 1 kHz. Use your oscilloscope to verify that this same voltage appears on the output of the IC.
6. Increase the signal level to 20 V peak-to-peak. Verify that the output voltage is the same.
7. Place a 1,000-Ω resistor between the output terminal and the common. The signal voltage should remain unchanged. If you can measure the output impedance, you will probably measure 10 Ω. This is the resistor we added in series with the output terminal.

there will be a dc path for base current in the input differential pair. Without this path the IC amplifier cannot function.

This circuit has a gain of 1, a high input impedance, and a low output impedance. These are the characteristics of a stacked emitter follower. This circuit is better than a stacked emitter follower; it is less expensive, smaller, and more accurate. This assumes that the output current and voltage are adequate.

Positive Gains Greater than 1

The circuit in Figure 5.2 can be modified to have any gain from 1 to about 100. Higher gains are possible depending on the type of IC amplifier being used. For the moment, a gain of 10 will demonstrate how the feedback circuit works.

In the circuit of Figure 5.4, an output voltage divider attenuates the output signal by a factor of 10. This voltage attenuator is called a *feedback divider*.

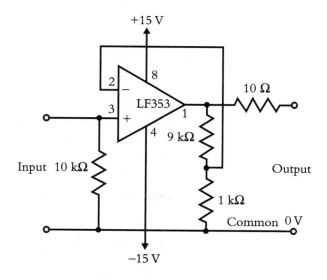

The feedback attenuator consists of a 1-kΩ and a 9-kΩ resistor. One-tenth of the output is fed back to the input. The gain is the reciprocal of this attenuation factor.

Figure 5.4 A positive gain 10 circuit using an IC amplifier

160 PRACTICAL ELECTRONICS

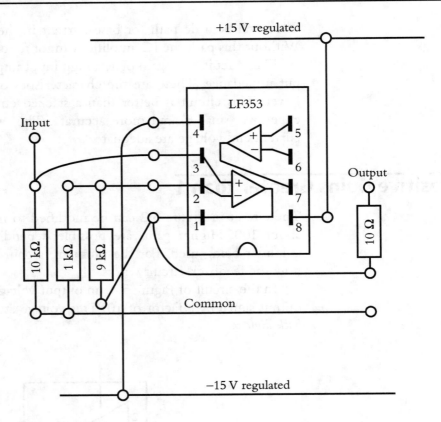

The 9-kΩ resistor can be formed from two 18-kΩ resistors in parallel. If pin 3 of the IC is left open (no 10-kΩ resistor to common or no input signal source), the circuit is very sensitive to noise and there is no path for input base current.

Figure 5.5 The construction of the circuit in Figure 5.4

When the circuit processes a signal, the output signal voltage must adjust until the two base voltages are equal. If one-tenth of the output is fed back to the negative input base, the output voltage must increase by a factor of 10 to cause equal signals to appear on the two bases. If the attenuation factor is 20:1, the gain will be 20. The rule to calculate this is as follows: Place an input signal on the input base. Calculate the output level that is required so that this same voltage will appear on the second or feedback base. The gain will be the reciprocal of the output attenuation.

The higher the gain, the larger the gain error. As an example, if the gain is 1, and the input signal is 1 V, the error signal is 1 µV. The ratio of input signal to error signal is 1,000,000:1. Assume the output signal is still 1 V. If the gain is 20, the input signal is 1/20 the previous value, or 0.05 V. The difference signal on the differential stage must still be 1 µV. Remember, the *open-circuit gain* is still 1,000,000. The ratio of input signal to error signal is now 50,000:1. This means that the signal gain error is one part in 50,000. This is still a small error, and it can be neglected. As you can see, excess gain (feedback factor) provides accuracy. By requiring a gain of 20, the excess gain has been reduced by a factor of 20 and the circuit is not as accurate. In a practical sense, the accuracy is limited by the resistor values and not by the feedback factor. This type

---

LEARNING CIRCUIT 34

Using Feedback and an IC Amplifier to Provide Positive Gain

You will need (in addition to your circuit board and measuring equipment):

1 LF353 IC amplifier

1 10-Ω, 1 1-kΩ, 1 9-kΩ (make from two 18-kΩ resistors in parallel), and 2 10-kΩ resistors

Add the output attenuator in Figure 5.4 to the circuit you built in Learning Circuit 33. Show that the gain is now +10. If your signal generator does not have an output level adjust, it may be necessary to reduce the generator output by using an attenuator. A series 10,000-Ω and a 100-Ω resistor can serve as a voltage divider to attenuate the generator signal for our lessons. The reduced input signal is taken across the 100-Ω resistor.

The gain of the circuit in Figures 5.4 and 5.5 can be changed to about 20 by paralleling the 1,000-Ω resistor with a second 1,000-Ω resistor. Observe this gain change using your oscilloscope. Be sure the output signal at a gain of 20 is kept less than 10 V peak. You may have to reduce the level of the input signal.

of feedback circuit cannot have a gain of less than +1. Of course, an input attenuator can always be added to limit the gain, but this solution has drawbacks, as the input impedance must be reduced to form a practical voltage divider.

Feedback and a Negative Gain

Gain is the ratio of change between an output signal and an input signal. Negative gain means that the output change is in the opposite direction. For example, if the output changes 2 V for a 1-mV input change, the gain is +2,000. If the output changes −2 V for an input change of 1 mV, the gain is −2,000.

To obtain negative gain using feedback, the positive input terminal is connected to the common. A feedback resistor is placed from the output terminal 1 to the negative input terminal 2. A second input resistor is connected from the negative input terminal 2 to the input of the circuit. This circuit is shown in Figures 5.6 and 5.7. The negative gain amplifier responds when a voltage is placed between the input and the common. As we saw in the last section the largest signal difference that is permitted in the differential input stage is 10 µV. Since the positive input terminal is at 0 V, the negative input terminal can only move ±10 µV. When gain is

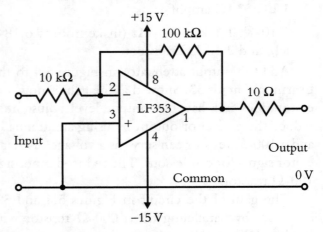

A negative gain 10 amplifier using an IC amplifier.
The gain is the ratio of 100 kΩ to 10 kΩ.

Figure 5.6 A negative gain feedback amplifier

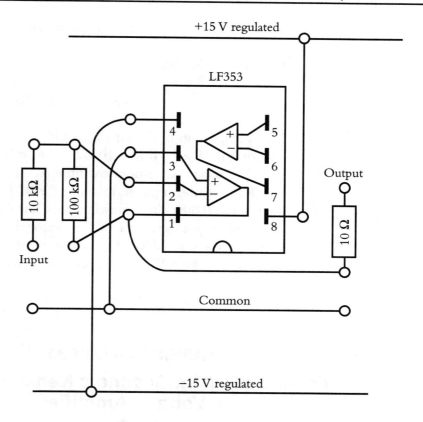

Figure 5.7 The construction of the circuit in Figure 5.6

calculated, it is safe to assume that this point does not move. For this reason, this point is called a *virtual ground*. It is also called the *summing point*. This type of gain control is called *operational feedback*. The feedback resistor is R_2 and the input resistor (which is also a feedback resistor) is R_1. When a signal voltage is placed on R_1, current flows to the negative input terminal of the amplifier. Since this point cannot move more than 10 μV, the input current is known. The input current is:

$$I_{IN} = V_{IN}/R_1 \qquad (5.1)$$

Where does this current go? There is no path into the base of the input transistor, so it must flow in resistor R_2. In fact, the output of the amplifier will adjust so that the input voltage is essentially 0. The output voltage is equal to the input current times the feedback resistor with a minus sign. This means that:

$$V_{OUT} = -IR_2 \tag{5.2}$$

But I equals V_{IN}/R_1. Therefore:

$$V_{OUT} = -V_{IN} \times R_2/R_1 \tag{5.3}$$

The voltage gain of the amplifier with feedback is the negative ratio of resistors, or $-R_2/R_1$. If $R_1 = 10$ kΩ and $R_2 = 100$ kΩ, the gain will be -10.

The voltage at the negative input terminal of the IC is the error signal. The negative input terminal to the IC is the summing point. If the gain is 1,000,000, this signal is always less than 10 µV. If the gain is 1, the largest input signal is 10 V. The error signal at the summing point is 10 µV. If the gain is 100, the largest input signal is 0.1 V and the error signal is still 10 µV. For this smaller signal, this error is one part in

---

LEARNING CIRCUIT 35

Constructing and Testing a Negative Gain IC Voltage Amplifier

You will need (in addition to your circuit board and measuring equipment):

1 LF353 IC amplifier

1 10-Ω, 1 10-kΩ, and 1 100-kΩ resistor

1 1.5-V battery

1. Modify the circuit board using the feedback circuit shown in Figures 5.6 and 5.7. Set $R_2 = 100$ kΩ and $R_1 = 10$ kΩ. Connect a sine wave signal at 0.5 V 1 kHz between the input terminal and the signal common. Adjust this signal level until the output signal of the amplifier is 5 V. Now change the signal to a square wave and note the same gain. Use a load resistor of 1 kΩ.
2. Set $R_2 = 10$ kΩ. Place a 1.5-V dry-cell battery in series with the signal generator lead. Note that the output signal is offset by -1.5 V. Reverse the polarity of the battery and note the new offset.

10,000. Just like the circuit using potentiometric feedback, the percentage error depends on the excess gain or feedback factor. For all practical purposes, the summing point does not move. This is why it is called a virtual ground. Later we will see why the error voltage increases at higher frequencies. You may try to look at the summing point with your oscilloscope, but you will be disappointed. You will add noise to the circuit, and the signal will be too small to observe.

In potentiometric feedback, the excess gain kept the two input bases at the same signal potential. In the operational feedback circuit, the same thing is true except that one of the bases is connected to 0 V. The excess gain requires that the negative input terminal stay near 0 V.

Operational feedback circuits can have a negative gain of less than 1. If $R_2 = 10$ kΩ and R_1 is 100 kΩ, the gain will be -0.1.

The input impedance of an operational feedback amplifier is the value of the input feedback resistor R_1. It is desirable in most designs to use an input resistor that is greater than 10 kΩ. At a gain of 100, the feedback resistor would have to be 1 MΩ. It is good practice to avoid using feedback resistors that are greater than 1 MΩ, because the circuits become noisy. Using a voltage divider in the feedback path can solve this problem. This circuit is shown in Figure 5.8.

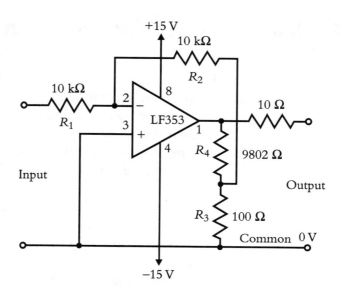

Figure 5.8 Operational feedback using a voltage divider

The values of the various feedback resistors can be determined as follows: Suppose the gain is to be −100. Select an arbitrary input voltage such as 0.1 V. The output voltage V_{OUT} equals −10 V. The input current is $I_{IN} = 0.1/R_1$. The current in R_2 must equal the input current. This defines the voltage V_2. Arbitrarily pick a value of R_3. The current in R_3 is $I_3 = V_2/R_3$. The output of the amplifier must supply both the feedback current and the current into R_3. The voltage across R_4 is $(V_{OUT} - V_2)$. The current in R_4 is $(I_{IN} + I_3)$. The resistor R_4 must equal $(V_{OUT} - V_2)/(I_{IN} + I_3)$.

To work an example, let R_1 and R_2 be 10 kΩ. Assume an input voltage of 0.1 V. The input current is 0.1/10,000 = 0.01 mA. The voltage $V_2 = -0.1$ V. Select $R_3 = 100$ Ω. The current in R_3 is 1 mA. The voltage across R_4 is (10 V − 0.1 V) = 9.9 V. The feedback current plus the current in R_3 is 1.01 mA. Using Ohm's law, $R_4 = 9.9/0.00101 = 9,802$ Ω. The circuit gain is 100 and the largest resistor is 10 kΩ.

Two integrated circuits can provide a positive gain less than unity (a gain less than 1). The first amplifier attenuates the signal and changes polarity. The second IC is a unity gain of −1. The combination provides a positive gain less than unity.

Using Feedback to Correct for Errors in the Signal Path

Feedback circuits are self-adjusting. The signal voltages in the circuit adjust until there is a balance at the error point or summing point. As an example, suppose a load resistor demands more output current. This greater current must flow in the internal resistance of the output circuit. This requires a larger drive signal. In a feedback circuit, the error signal increases by a small amount and supplies this extra signal. The result is that the output voltage does not appear to change. A signal source that does not change when there is a change in output current has a low output impedance. For most IC amplifier circuits, the output impedance is in the milliohm range. This is a very low value that is often hard to measure.

The stacked emitter follower circuit in Learning Circuit 25 had a problem: The diodes that determined the base voltages set the static current level in the two transistors. If the base voltage to emitter voltage is too low, the transistors will not conduct. In this case, the input signal has to be large enough to turn the transistor on before there can be an output signal. The resulting output signal for a sine wave input is shown in Figure 5.9.

This problem is avoided by placing the stacked emitter follower inside a feedback circuit. This circuit is shown in Figure 5.10.

The output signal will be a clean sine wave. The signal internal to the feedback structure can be nonlinear. The signal driving the emitter followers will exactly compensate for the diode problem. The output of the IC amplifier will have a steep waveform around the zero crossings that exactly corrects for the inability of the output stage to conduct current.

This circuit has the advantage of limiting power dissipation. When there is no signal, there is no current in the stacked emitter portion of the circuit. When there is a signal, the circuit will deliver this signal to the load without significant distortion.

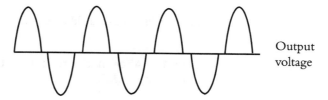

Output voltage

This voltage pattern occurs when a stacked emitter follower circuit does not conduct around 0 V. This problem results when the potential difference between the transistor bases is not great enough.

A small resistor in series with the diodes in the base circuit can reduce this effect. If both emitter followers conduct around zero output voltage, this pattern will not appear. Excess static current can result in transistor heating. Some distortion may be present even if current flows during the entire cycle.

Figure 5.9 The distorted voltage waveform for a stacked emitter follower

168 PRACTICAL ELECTRONICS

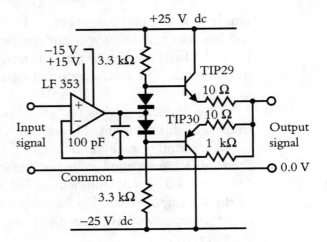

The feedback path is through the 1 kΩ resistor. The emitter followers have some delay (phase shift) and this can affect the stability of the circuit. The 100 pF capacitor makes sure that the circuit is stable by connecting the output of the IC to the input at high frequencies.

Placing the emitter followers in a feedback circuit removes any distortion caused by biasing.

You may build this circuit although it is not one of the Learning Circuits.

Figure 5.10 A feedback circuit using the stacked emitter followers as an output circuit

Feedback Correction of Internal Distortion

The way feedback corrects for internal distortion can be shown by modifying the circuit shown in Figure 5.2. Normally the feedback is taken from the output terminal of the IC amplifier. As an experiment, we are going to intentionally distort the gain by placing back-to-back parallel signal diodes in series ahead of the load resistor. This circuit is shown in Figures 5.11 and 5.12. A distorted sine wave with a horizontal flat segment around each zero crossing occurs at terminal B. If the feedback is taken from the load resistor with the diodes in place, the distortion is removed. The diodes are now inside the feedback loop.

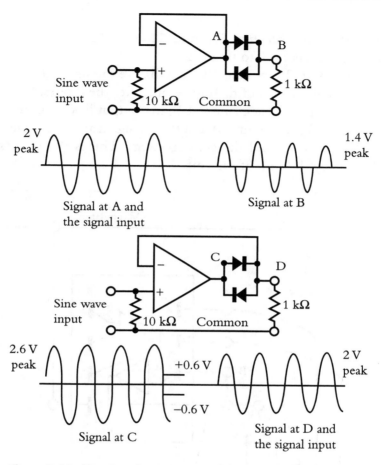

Figure 5.11 The signal patterns in a feedback circuit with added diodes

---

⊸⟋⊸ LEARNING CIRCUIT 36 ⊸⟋⊸
Using Feedback to Correct for Signal Distortion

Your will need (in addition to your circuit board and measuring equipment):

    1 LF353 IC amplifier
    2 1N4148 signal diodes

(Continued)

1. Build the circuit shown in Figures 5.11 and 5.12.
2. Connect the feedback to point A.
3. Set the signal to 2 V peak at 1 kHz. Observe the signal at the output of the amplifier ahead of the diodes and after the diodes.
4. Reconnect the feedback to point B, the output load resistor.
5. Observe the signal at points A and B. The signal at A will have a very steep transition around the zero crossings. This is the error signal multiplied by the gain of the amplifier.

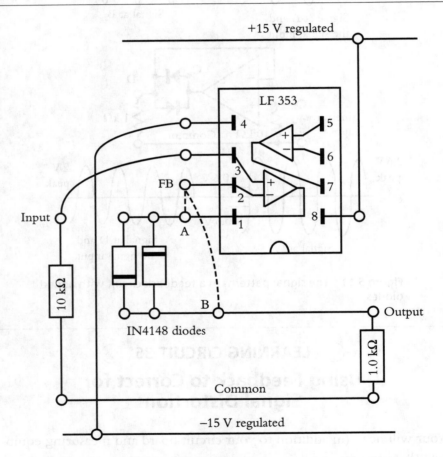

Connect FB (pin 2 of the IC) to A or B. In position B the diodes are inside the feedback loop.

Figure 5.12 The construction of the circuit in Figure 5.11

Feedback and Stability

Negative feedback requires the subtraction of signals at the input. In a feedback circuit, the signals of interest must pass through gain stages (forward gain) and then return to the input. There is an issue that we have been ignoring up to this point: the phase shift and time delay inherent in all circuits.

When we studied passive circuits, we used a pointer system to show the effect capacitance had in delaying a sine wave. We also saw that sine waves that are shifted in phase do not add and subtract directly. All circuits have small capacitances that limit their amplitude response as a function of frequency. This means that phase shift and delay are a part of every circuit design. A feedback signal cannot subtract from an input signal unless the two signals are in phase. Even a few degrees of phase shift creates a problem.

The open-loop gain (gain without feedback) of an IC amplifier is very carefully controlled. The phase shift for open-loop gains greater than unity is limited to 90°. If the closed-loop phase shift ever reaches 180° and there is still gain, the circuit with feedback can oscillate in some configurations. As you will recall, phase shift is proportional to attenuation factor. If the amplitude decreases proportional to frequency, the phase shift is 90°. If an amplifier has an open loop gain of 1,000,000, it takes six decades of frequency for the gain to reach 1.

To control and limit the phase shift, the open-loop gain of the amplifier must begin to lose gain at a very low frequency. As an example, suppose the gain is 1,000,000 as it starts down at 10 Hz. The gain will be 100,000 at 100 Hz and 10,000 at 1 kHz. To maintain this same slope, the gain reaches unity at 10 MHz. This does not mean that the amplifier can handle 10-MHz sine waves. It only means that the open-loop phase character is controlled to this frequency. An amplifier with this controlled phase shift is called an *internally compensated amplifier.* Fortunately for the user, this difficult problem has been solved by the IC designer.

In the previous example, if the feedback resistors are set to provide a closed-loop gain of 100, the amplifier has no excess gain above a frequency of 100 kHz. At 10 kHz, the excess gain is 10, and at 1 kHz, the excess gain is approximately 100. At 1 kHz the error signal must be equal to the output signal divided by 100.

The bandwidth (the highest response frequency) of an IC amplifier depends on the closed-loop gain. Gain is obtained at the expense of bandwidth. This is one reason why an individual IC amplifier should not be used to supply all the gain. If more gain is required, then two IC amplifiers can be cascaded (placed in series).

The open-loop phase shift of a compensated amplifier is limited to 90°. (Designers use a few tricks to increase this phase shift over a limited part of the spectrum.) A phase shift of 180° in a negative gain amplifier results in an output signal that is in phase with the input signal. When this signal is fed back, it is called *positive feedback*.

To understand positive feedback, imagine that if your car was going faster than you wanted, you responded by *increasing* rather than decreasing its speed. The car would then go faster and faster until it broke down or ran out of gas. If the car was going slower than you wanted, positive feedback would be required to further decrease the speed until the car stopped. In electronic circuits, positive feedback generally results in an oscillation. The circuit goes in one direction until it can't go any farther, and there is a sharp change to the opposite direction—this is the oscillation. In circuits with too much phase shift, there may be no oscillation, but the circuit might border on instability. An unstable response is shown in Figure 5.13.

If you add a capacitive load to the output of an IC amplifier (e.g., as a shielded output cable such as an oscilloscope probe), it can disturb the internal compensation. This capacitance adds phase shift to the open-loop response, and the total phase shift can exceed 180°. The problem is more apt to appear when the closed-loop gain is unity. This chance of instability is the reason for the 10-Ω resistor added to the output of the feedback circuits. This resistor provides an additional margin of safety.

The phase shift in a closed-loop feedback system is 90° divided by the feedback factor. At the -3-dB frequency there is no more feedback (no excess gain), and the phase shift approaches 90°. Above this frequency there is no excess gain and the phase shift can continue to increase. As an example, at a lower frequency where there is a feedback factor of 100, a phase shift of 90° would be reduced to 0.9°.

Instability problems are most apt to occur when the bandwidth and the feedback factors are high. If there is an instability, the oscillation can be low-amplitude at a frequency above 10 MHz, which might go unnoticed. It is always a good idea to test a circuit for stability using a square wave voltage. The circuit should be able to accept a capacitance load and not show signs of oscillation. To check for stability, it is a good idea to try several values of capacitance loading from 100 pF to 0.01 µF.

Feedback and IC Amplifiers 173

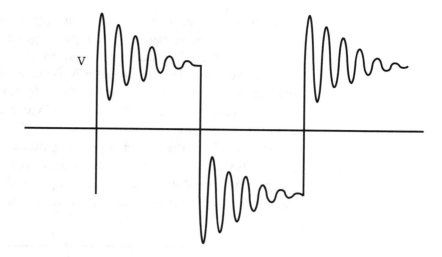

Figure 5.13 A response to a square wave voltage indicating a very unstable condition

When a circuit is close to being unstable, there is "ringing" (the oscillation shown in Figure 5.13) on the leading edge of the output square wave. Designs with ringing are undesirable. A change in temperature or a change in load could take the circuit over the edge.

Testing for instability should involve full-scale signals as well as small signals. Large signals can sometimes operate the circuit where the open-loop gain is lower. This reduces the feedback factor and gives the appearance of stability. Small testing signals are more apt to show any instability. A better test is to superpose a small square wave signal on a large sine wave signal. This provides the most visibility for problems. The large signal requires the full range of output operating points, and the small signal indicates any sign of instability over the operating range.

Single Supply Operation

In the circuits we have discussed so far, the power supplies have been symmetrical about a midpoint. This midpoint has been associated with the input circuit common and with the feedback design. Remember, this midpoint is not connected to the IC amplifier. In effect, the amplifier operates from a single potential difference.

The voltages on the input terminals of an IC amplifier must be in the range of the power supply. Further, the input voltages must not approach the power supply voltages or the input stage cannot function. The manufacturer of the IC amplifier is careful to specify these voltage limits.

One way to solve this problem is to use a voltage divider on the positive input to the IC amplifier. Feedback forces the negative input to be near this same potential. The voltage divider establishes a common or zero for the circuit.

If one side of the power supply is grounded, the voltage divider can still be used. The potential at the positive input is the reference potential for the amplifier. In effect, the input and output signals are referenced to this potential. If the signals originate referenced to the power

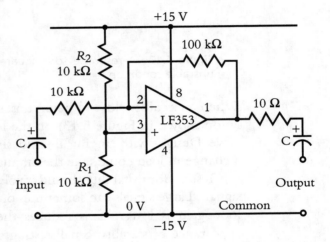

This amplifier provides gain around a point that is 7.5 V above the common. The capacitors allow ac signals to be amplified with respect to the common. At ac the gain is 10. The −3 dB points are determined by the values of C. At the input this occurs at the frequency where the reactance of the capacitor equals 10 kΩ. At the output the capacitor value depends on the load resistor.

A capacitor is often placed across R_1. Another solution is to replace R_1 with a 7.5 V zener diode. If this is done, then R_2 should be reduced in value to allow zener current to flow.

Figure 5.14 An amplifier circuit using a single supply voltage

supply common, then a coupling capacitor can be used to block any offset. A typical circuit is shown in 5.14.

The Differential Amplifier

Noise and interference problems are often encountered in interconnecting signals from different devices. The problem often appears as a difference in potential between signal commons. Often the interconnecting signal cable will provide a path for unwanted current flow. If this current flow adds to the signal, the resulting interference can be objectionable.

A differential amplifier can limit this current flow and at the same time reject the difference in potential between reference conductors. This circuit is shown in Figures 5.15 and 5.16.

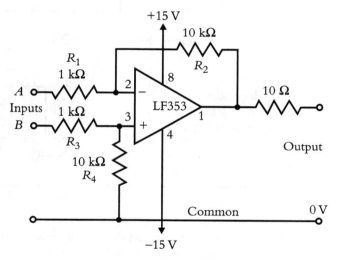

Signals A and B are potentials with respect to the common. This amplifier has a gain of 10 to the difference between A and B. The output equals $10(B - A)$. If $A = 0$, the output equals $10B$. If $B = 0$, the output equals $-10A$. If $A = B$, the output is 0.

If A is connected to B, then by definition $A = B$ and the output is 0. This is called *common-mode rejection*. The common mode signal is rejected if the ratio $R_2/R_2 = R_4/R_3$.

If A is a 0.1 V sine wave around 5 V dc and B is a -0.1-V sine wave around 5 V dc, the output is a 2-V sine wave around the common as the 5 V dc is a common-mode signal and it is rejected.

Figure 5.15 A differential amplifier

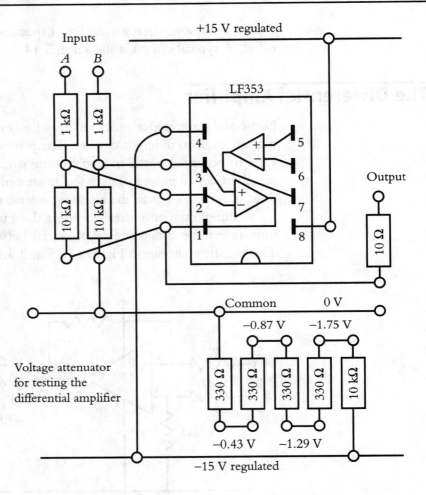

Use clip leads to connect the input terminals to the various points on the voltage attenuator. Each 330-Ω resistor has a 0.43-V voltage drop. The amplifier output should be 4.3 V when the input leads are across any 330-Ω resistor. Reverse the input leads and the output voltage will change sign. Test across two resistors.

Figure 5.16 The construction of the circuit in Figure 5.15

This circuit has four feedback resistors. The resistor values are selected so that $R_1/R_2 = R_3/R_4$. In the circuit analysis we will use, all the signal voltages are referenced to the power supply common. The gain to B is R_2/R_1. This requires the signal A to be 0 V. The gain to A is $-R_2/R_1$. If $A = B$, then the two signals cancel and the output signal is 0.

LEARNING CIRCUIT 37
Building and Applying a Differential Amplifier

You will need (in addition to your circuit board and measuring equipment):

    1 LF353 IC amplifier

    1 10-Ω, 4 330-Ω, 2 1-kΩ, and 3 10-kΩ resistors

1. Build the circuit shown in Figures 5.15 and 5.16. Use $R_2 = R_4 = 10$ kΩ and $R_1 = R_3 = 1$ kΩ.
2. Build a voltage divider across the power supply to obtain several voltages. The resistors can be 10 kΩ and four 330 Ω. The voltage across each 330-Ω resistor is about 0.437 V.
3. Connect the IC amplifier input terminals across any one of the 330-Ω resistors. The gain is 10, so the output voltage should be 4.4 V. You should get the same result regardless of which resistor you select.
4. Reverse the leads and notice that the output voltage reverses polarity. Leave the input terminals connected to one of the bottom three 330-Ω resistors.
5. Place the signal generator with a 1-V sine wave at 100 Hz across the top 330-Ω resistor. Notice that this signal does not appear in the output. This signal is added to both inputs and is rejected.
6. Turn off the power supply. The gain of the amplifier is 10.
7. Switch resistors R_1 and R_2.
8. Switch resistors R_3 and R_4. The gain should now be 0.1.
9. Turn on the power supply.
10. Place the input leads across one of the power supply diodes. Observe the signal at the output of the amplifier. This signal is the voltage across the diode attenuated by a factor of 10, referenced to the common conductor of the power supply. Do not attempt to measure the diode voltage by placing the oscilloscope across the diode. The generator and the oscilloscope can both be safely grounded, and this will short out the transformer secondary winding. This ability to look at signals anywhere in a circuit is a valuable asset.

There is only gain to the difference between A and B. This gain is the ratio of R_2/R_1. The amplifier responds to the difference signal and ignores the average signal. This is the meaning of the term *differential amplifier*. If the resistors are matched to 1% resistors, the average value will be rejected to within 1%. The rejection of the average value is called *common-mode rejection*.

This circuit can be used to measure the voltage across any circuit element, including the diodes in the rectifier circuit. Here the common-mode voltage can be 50 V peak. To use the amplifier for this application, the gain must be less than 1. If $R_2/R_1 = 0.10$, the differential amplifier can accommodate this 50-V signal.

Active Low-Pass Filters

The IC amplifier can be used to provide low-pass or high-pass filtering. In Figure 1.18 we saw the amplitude and phase response of an RC low-

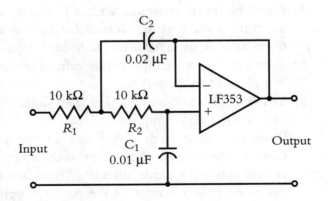

This low-pass filter has a −3-dB cutoff frequency at 1.1 kHz.
If the resistors are doubled, the cutoff frequency halves.
If the capacitors double, the cutoff frequency halves.

Figure 5.17 An active second-order low-pass filter

Feedback and IC Amplifiers   179

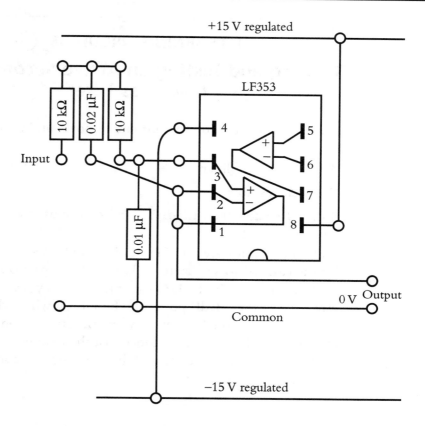

The 0.02-μF capacitor can be made from two 0.01-μF capacitors in parallel.

Figure 5.18 The construction of the circuit in Figure 5.17

pass filter. This type of filtering is often used in attenuating high-frequency interference. In some applications it is desirable to have more attenuation above the cutoff frequency.

The circuit in Figures 5.17 and 5.18 provides an attenuation slope that is proportional to the square of frequency. Stated another way, the attenuation slope is 40 dB per decade. At 10 times the frequency, the attenuation factor is 100. This is called a *second-order filter*.

LEARNING CIRCUIT 38

Building and Testing an Active Second-Order Low-Pass Filter

You will need (in addition to your circuit board and measuring equipment):

1 LF353 IC amplifier

2 10-kΩ resistors

1 0.01-μF and 1 0.02-μF capacitor (you can use two 0.01 μF capacitors in parallel)

1. Build the circuit shown in Figures 5.17 and 5.18.
2. Set the two resistors equal to 10 kΩ and the capacitors equal to $C_2 = 0.02$ μF and $C_1 = 0.01$ μF. Check the frequency response using sine waves. Note the -3-dB point, the frequency where the attenuation factor is 0.1, and the frequency where the attenuation factor is 0.01.
3. Double the value of C_2 and notice that the frequency response has a significant peak. Also notice that the overshoot for a square wave has increased.

This filter has the same characteristics as the RLC circuit described in Figure 2.9. This is an example of where an active circuit has the same response as a passive circuit using an inductor. However, this circuit can be made to operate at frequencies as low as 1 Hz, which is almost impossible with inductors. Inductors saturate, and they have their own natural frequency. They are bulky and expensive. The circuit approach has the further advantage of providing a low output impedance.

The circuit can be designed to have different cutoff frequencies and different amplitude responses near the cutoff frequency. The exact -3-dB frequency depends on the ratios between capacitors and resistors. A good approach in design is to set the two resistors equal. When $C_2/C_1 = 2$ and $R_1 = R_2$, the frequency response is near optimum and the square wave response will have a 7% overshoot. If $C_2/C_1 = 1$, the square-wave response will have no overshoot. The −3-dB point will be approximately

$$f = 1/(2\pi R \sqrt{C_1 C_2}) \qquad (5.4)$$

A fourth-order filter can be built by cascading two second-order filters. The amplitude response will fall off proportional to the fourth power of frequency. The terminal slope will be 80 dB per decade. The individual second-order characteristics can be adjusted to provide optimum flat frequency response, minimum square wave overshoot, or a very sharp knee. These are all topics in filter design taught to engineers.

Active High-Pass Filters

When the capacitors and resistors in Figure 5.17 are reversed, a high-pass filter is formed. The filter is optimum when the capacitors are

---

⌇ LEARNING CIRCUIT 39 ⌇
Building and Testing a High-Pass Circuit

Your will need (in addition to your circuit board and measuring equipment):

1 LF353 IC amplifier

2 0.01-μF capacitors

1 10-kΩ and 1 20-kΩ resistor (you can use two 10-kΩ resistors in series)

1. Build the circuit shown in Figures 5.19 and 5.20.
2. Set the capacitors to 0.01 μF and the resistors to $R_1 = 10$ kΩ and $R_2 = 20$ kΩ. Measure the frequency response and note the −3-dB point. It should be very close to the same cutoff frequency in the previous

(Continued)

182 PRACTICAL ELECTRONICS

> learning circuit. Note the frequencies where the signal is attenuated by a factor of 10 and a factor of 100.
> 3. Test the circuit using a square wave signal. Note the response waveform. When the input waveform goes positive, the output follows and immediately returns to 0 overshooting 0. The same thing happens when the input wave goes negative. These undershoots and overshoots are characteristic of high-pass filters. A high-pass filter cannot pass dc, so the average voltage in the output must be 0.

equal and the resistors have a ratio of 2:1. The terminal slope for this filter is 40 dB per decade. This filter is shown in Figures 5.19 and 5.20.

The cutoff frequency for this filter is given by the equation

$$f = 1/(2\pi C\sqrt{R_1 R_2}) \qquad (5.5)$$

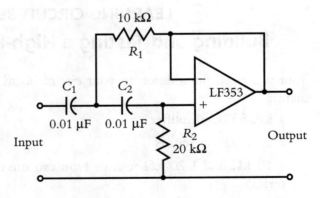

This high-pass filter has a −3-dB point at 1.1 kHz. If the resistors are doubled, the cutoff frequency halves. If the capacitors are doubled, the cutoff frequency halves.

Figure 5.19 An active second-order high-pass filter

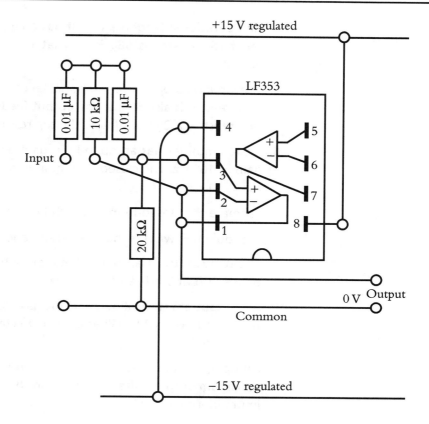

The 20-kΩ resistor can be made from two 10-kΩ resistors in series.

Figure 5.20 The construction of the circuit in Figure 5.19

SELF-TEST

1. An IC amplifier has an open-loop gain of 200,000. The closed-loop gain is +1. What is the maximum error signal at dc for 10-V output?

2. In a potentiometric feedback circuit, the output feedback attenuator is 3,000 Ω and 1,000 Ω. What is the closed-loop gain? What is the gain if the resistors are reversed?

3. In an IC amplifier, the output is connected to the negative input. What is the gain from the positive input terminal?

4. An operational feedback circuit has an input resistor of 200 kΩ. The feedback resistor is 560 kΩ. What is the gain? Indicate the gain polarity.

5. The input resistor to an operational feedback circuit is 50 kΩ. The feedback resistor is also 50 kΩ. The output feedback attenuator has a bottom resistor of 200 Ω. What is the top resistor if the gain is to be −30?

6. An IC amplifier has an open-loop gain of 100,000 that starts losing gain at 100 Hz. The closed-loop gain is 20. What is the feedback factor at 10 kHz?

7. Estimate the closed-loop phase shift in problem 6 at 10 kHz.

8. In problem 6, what is the estimated bandwidth?

9. The bandwidth of an operational amplifier is 50 kHz. Estimate the phase shift at 5 kHz and at 50 kHz.

10. An output emitter follower circuit has 3% distortion. The feedback factor at 1 kHz is 200. What is the expected distortion in the output waveform?

11. An operational feedback circuit has an input 100-kΩ resistor. If the maximum output voltage is 10 V and the gain is 10, what is the maximum input current?

12. A differential circuit uses four equal 20-kΩ resistors. What is the gain to the difference signal? What is the common-mode gain?

13. A low-pass second-order filter uses 20-kΩ resistors. What are the capacitor values if the −3-dB frequency response is to be at 2 kHz?

14. In problem 13, what are the capacitor values if the −3-dB frequency is to be 20 kHz?

15. A high-pass second-order filter uses 0.1-μF capacitors. What are the resistor values if the −3-dB frequency is to be 20 Hz?

ANSWERS

1. 10 V/200,000 = 50 μV.

2. The attenuation is 1,000/4,000 and the gain is 4. For reversed resistors, the gain is 4/3.

3. 1.0.

4. The gain is −2.8.

5. Assume the input is 0.1 V. The output voltage is 3 V. The voltage at the junction of the feedback attenuator is −0.3 V. The voltage across the top resistor is 2.9 V. The current in the 200-Ω resistor is 0.1/200 = 0.5 mA. The current in the 50-kΩ resistor is 0.1/50,000 = 2 µA. The total current in the top resistor is 502 µA. By Ohm's law, the resistor is 2.9/0.000502 = 5.776 kΩ.

6. The ratio of 10 kHz to 100 Hz is 100. The open-loop gain is only 1,000. If the closed-loop gain is 20, the feedback factor is 50.

7. The phase shift is 90°/50 = 1.8°.

8. The bandwidth is 50 × 10 kHz = 500 kHz.

9. The phase shift at 50 kHz is approximately 90° and at 5 kHz, 9°.

10. 3%/200 = 0.015%.

11. The maximum input voltage is 1.0 V. The input resistance is 100 kΩ. The maximum input current is 10 µA.

12. The gain to difference signals is 1. The gain to common-mode signals is 0. It does depend on accuracy of the resistors.

13. The capacitors are both 0.004 µF.

14. At 20 kHz the nominal capacitor values are 0.0004 µF. To allow the ratio of capacitors to be 2:1, multiply C_2 by $\sqrt{2}$ and C_1 by $1/\sqrt{2}$. The result is C_1 = 0.00283 µF and C_2 = 0.00586 µF.

15. The nominal resistor values are 79.6 kΩ. To allow the ratio of resistors to be 2:1, multiply R_2 by $\sqrt{2}$ and multiply R_1 by $1/\sqrt{2}$. The result is R_1 = 56.25 kΩ and R_2 = 112.58 kΩ.

6 IC Applications

Objectives

In this chapter you will learn:

- about voltage and current regulators
- how a comparator circuit functions
- how to generate a sawtooth waveform
- about bistable and astable multivibrator circuits
- the principles of A/D and D/A converters

Integrated circuits are used to perform many tasks besides the one we looked at in chapter 5, providing voltage gain. They are used in waveform generation, timing circuits, voltage regulators, signal measurement, oscillators, comparator circuits, switching, and buffering, to mention just a few applications. Units with high-frequency performance are used to process signals in radio and television. In fact, the majority of today's electronic designs make use of integrated circuits. In many mass-produced products, several circuit functions are even combined into one IC. In this chapter we will examine several of these applications in more detail.

Voltage Regulators

Three-terminal IC voltage regulators are the simplest way to obtain a regulated voltage for a circuit. Components are available that provide standard plus or minus voltages. Typical voltages are 5 V, 10 V, and 15 V. These ICs provide regulation plus overload and overtemperature protection, and have their own internal reference zener diodes. They provide excellent performance and are quite small and inexpensive. The user must provide an electrolytic capacitor from the output to the common to guarantee dynamic stability. Versions are available that allow the user to set the regulated voltage by adding two external resistors. You may wish to buy 15-V three-terminal regulators and use them in your circuit board, although this is not required.

To learn about voltage regulators, you are going to build one. A typical positive 5-V regulator circuit is shown in Figure 6.1. This circuit is far more stable than the emitter follower circuits you built in chapter 4. The output voltage adjusts so that the signals on the inputs of the IC are equal. The voltage divider determines the regulated voltage. A negative power supply is unnecessary because all of the operating voltages of the IC are positive. This circuit should always have an electrolytic capacitor connected from the regulated output to the common. A typical value might be 10 μF. This is also required for a three-terminal regulator.

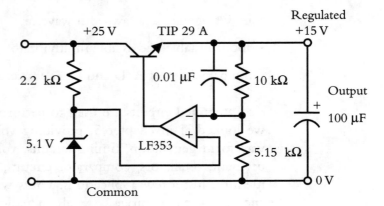

5.15 kΩ = approximately 4.7 kΩ in series with 470 Ω.

Figure 6.1 A positive voltage regulator using an IC amplifier

IC Applications   189

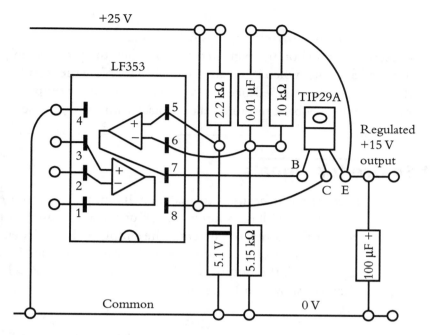

The 5.15-kΩ resistor can be made from 4.7 kΩ and 470 Ω in series.

You may want to substitute this regulator for the one you already have. It is a better circuit.

Figure 6.2 The construction of the circuit in Figure 6.1

 **LEARNING CIRCUIT 40**
Building and Testing a Regulated Power Supply

You will need (in addition to your circuit board and measuring equipment):

    1 LF353 IC amplifier

    1 TIP29A transistor

    1 5.1-V zener diode

    1 2.2-kΩ, 1 4.7-kΩ, 1 470-Ω, 1 10-kΩ, and 1 100-Ω resistor

    1 0.01-μF and 1 100-μF capacitor

(Continued)

1. Build the circuit shown in Figures 6.1 and 6.2.
2. Measure the ripple on the regulated output when the power supply is unloaded and then when it is loaded with 100 Ω. Note the regulation.
3. Measure the output impedance. Be careful to measure the output voltage where the feedback resistors make contact with the output lead.

A Current Source for Signals

Feedback can be used to make an IC amplifier into a current source instead of a voltage source. An input voltage is used to provide a proportional amount of current. The output voltage depends on the load resistor. For example, if a feedback circuit provides 1 mA per volt of input signal, the output is 2 V if the load resistor is 2 kΩ. This feedback circuit is shown in Figure 6.3.

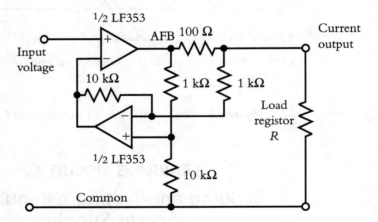

The power supply voltages for the IC amplifier are plus and minus 15 V. These are not shown on the schematic.

This circuit will not work without a load resistor. If the input is 1 V, volts the current output is 10 mA. With a load resistor of 500-Ω, the output voltage is 5 V. If the input is -1 V, the output current is -10 mA. The output voltage is then -5 V into 500 Ω.

The current output is 10 mA/volt of input.

Figure 6.3 A current source

The current sensing resistor is R_1. The unity gain differential stage measures the voltage drop across this resistor and feeds this signal back to the negative input. This is an example in which the feedback path has an active circuit. This circuit will overload without an output load resistor. If the sensing resistor is 100 Ω and the input voltage is 1 V, the output current will be 10 mA. If the load resistor is 500 Ω, the output voltage will be −5 V. The output voltage is −6 V if the load resistor is 600 Ω. This circuit provides a current source for ac and dc signals.

A constant current source that is often used in industrial measurement is shown in Figure 6.4. A sensor that measures pressure or flow has an internal current amplifier. The load resistor for the output of this amplifier is placed at the end of the signal loop, which might be hundreds of feet away.

The load current must flow in resistor R_{FB}. This is the current sensing resistor. The feedback circuit requires that the voltage across R_{FB} equal the input voltage. This means that $I_{OUT} = V_{IN}/R_{FB}$. The voltage across the load resistor changes with resistance value. As an example, assume that $R_{FB} = 100$ Ω. If the input voltage is 1 V, the current in the output is 10 mA. A load resistor of 500 Ω will have 5 V across its terminals. If the load resistor is 400 Ω, the voltage is only 4 V. The voltage drop across the connecting wires is ignored.

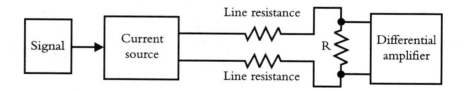

R is the sensing resistor. The signal across R is independent of the line resistance.

The signal can be associated with a grounded circuit or common at the signal source. The differential amplifier can be associated with a second grounded circuit. The potential difference between the grounds is rejected as common-mode signal.

Figure 6.4 A constant current loop used in measurement

A differential amplifier that is associated with a second signal reference common (ground) senses the voltage at the load resistor. The ground differential in potential between the sensing circuit common and the output reference common can be several volts. This potential difference is a common-mode voltage. The differential amplifier we discussed in the last chapter is used to reject this common-mode signal.

The Integrator

A constant current flowing into a capacitor causes the voltage to increase at a constant rate. This is analogous to putting money into the bank at one cent per second, so that the money in the account increases at a steady rate. These are two examples of *integration*. The voltage on the capacitor is the integral of current flow, and the balance in the account is the integral of money flow.

In an operational feedback circuit, the summing point does not move. If the input voltage is constant, the current to the summing point is also constant. If the feedback element is a capacitor, this constant current must flow into the capacitor. The result is a steady increase in voltage across the capacitor. This voltage is also the output voltage. The output voltage is the integral of the input voltage. This integrator circuit is shown in Figure 6.5.

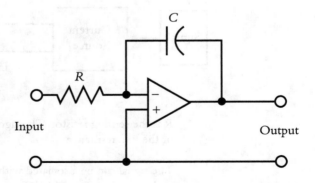

The power supply voltages are not shown.

Figure 6.5 A voltage integrator circuit

The output of the integrator circuit inverts the sign of the integral. To see the uninverted integral, a second unity gain inverting amplifier can be placed after the integrator. It is helpful to see graphically what the ideal integral is for various input voltage waveforms. At $t = 0$, the capacitor is assumed to have no charge on its plates. The charge is added or subtracted by the feedback amplifier. These integrals are shown in Figure 6.6.

Consider an input voltage that is 1 V for 0.1 second. Assume the output voltage is a ramp that goes minus to −1.0 V. If the input voltage then returns to 0 V, the output must stay at −1 V. Will the integrator hold the voltage for a minute or an hour? In practice, an IC integrator will begin to drift away from its proper value after a few seconds.

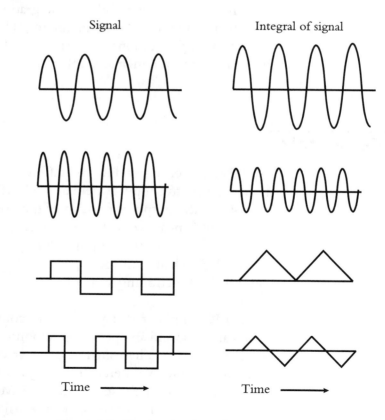

Figure 6.6 The integration of various waveforms

An explanation that provides some insight into the problem involves the circuit response at low frequencies. Assume the closed-loop gain at 100 Hz is 10 and the reactance of the capacitor is 100 kΩ. At 10 Hz the reactance is 1 MΩ and the closed-loop gain of the circuit is 100. Consider a frequency as low as 0.1 Hz. The closed-loop gain is 10,000 and the reactance is 100 MΩ. It is easy to see that as the frequency gets lower and lower, the closed-loop gain will eventually reach a limit. It cannot exceed the open-loop gain, and this is the point where there is no feedback. In other words, the integrator is no longer functional. Another thing happens: there is input base current. This current is no different from input signal current except it is much smaller. The result is an error in the integral that grows with time.

The integrator in Figure 6.5 is a valuable tool, but its use has a time constraint. Before the error becomes too large, the charge on the capacitor must be removed and the integration must start over. A digital up–down counter is also an integrator. The last count can be held in a register until the computer is turned off. This is one advantage a digital integrator has over an analog integrator.

The Comparator

An IC amplifier responds to the difference signal at its two inputs. Suppose the negative input is held at 4.5 V. If the positive input is greater than 4.5 V, the output is positive. If the positive input is less than 4.5 V, the output is negative. If there is no feedback and there is a great deal of open-loop gain, the output will respond with a full-scale signal for a very slight change in the input signal at right around 4.5 V. The IC amplifier is functioning as a comparator. This circuit is shown in Figure 6.7.

An IC amplifier that is internally compensated has a great deal of loop gain at dc and an open-loop response that starts to attenuate signals at perhaps 10 Hz. These are not the characteristics needed to perform a fast comparison. A different IC design is needed. The open-loop frequency response might start down at 1 MHz, not 10 Hz. Comparator amplifiers share some of the same internal construction as an IC amplifier, but they meet entirely different specifications.

IC Applications   195

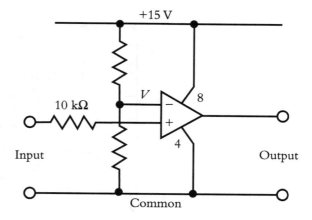

When the input voltage is at 0 V, the output is also near 0 V. If the input is greater than V, then the output voltage is near the positive power supply.

The polarity of the output can be reversed by placing the voltage attenuator on the positive input and the input signal on the negative input.

Figure 6.7 A comparator circuit using an IC amplifier

A Sawtooth Voltage Generator

A sawtooth voltage waveform is shown in Figure 6.8. The signal voltage starts at 0 V and ramps positive at a fixed rate. At a critical voltage, the voltage returns to zero and the voltage ramps positive, repeating the cycle. A negative going ramp with a rapid return to zero is also a

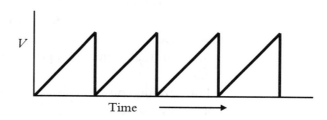

Figure 6.8 A sawtooth voltage waveform

196 PRACTICAL ELECTRONICS

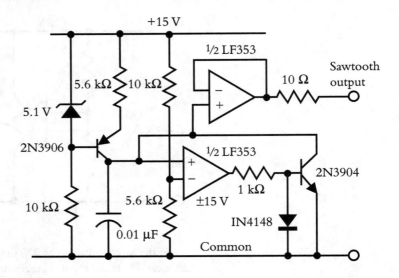

The 2N3906 transistor provides the constant current that charges the 0.01-μF capacitor. The bottom IC amplifier is a comparator. When the capacitor voltage reaches 5 V, the output of the comparator changes states to +15 V and turns on the transistor switch. (2N3904). This shorts out the capacitor, reducing the capacitor voltage to near 0. This changes the state of the comparator, allowing the capacitor to begin charging again. The upper IC amplifier is a buffer that isolates the capacitor and allows a user to place a load on the sawtooth signal.

Figure 6.9 A sawtooth voltage generator circuit

sawtooth waveform. A sawtooth voltage is used to deflect the electron beam in an oscilloscope. It is also used to move the electron beam on a television screen. In this case one sawtooth signal moves the beam across horizontally and another moves the beam vertically. The combination of two sawtooth signals generates a series of parallel lines called a *raster*.

A circuit that generates a positive sawtooth voltage is shown in Figure 6.9.

The circuit consists of a current source, a capacitor, a transistor clamp, a comparator circuit, and a buffer stage. The constant current is always flowing. This current flowing into the capacitor causes the volt-

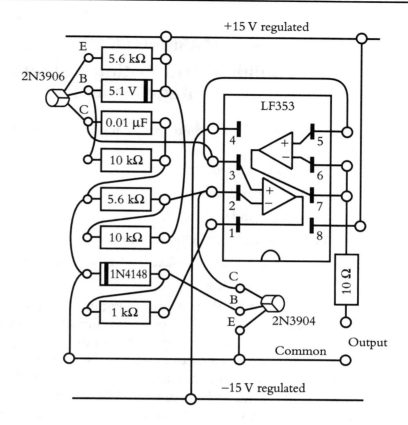

Figure 6.10 The construction of the circuit in Figure 6.9

age across the capacitor to ramp up. The positive input to the comparator is connected to the capacitor voltage. The negative input to the comparator is set to 5 V on a voltage divider. When the capacitor voltage reaches 5 V, the comparator output voltage changes from a negative voltage to the plus power supply voltage. The output of the comparator is connected to the base of the clamping transistor through a series resistor. When the clamping transistor is turned on, it discharges the capacitor. This of course resets the comparator, which turns off the transistor, and the capacitor again ramps up in voltage. The steady rise in voltage followed by a rapid discharge is the sawtooth waveform we are looking for.

LEARNING CIRCUIT 41
Building and Testing a Sawtooth Voltage Generator

You will need (in addition to your circuit board and measuring equipment):

1 LF353 IC amplifier

1 5.1-V zener diode

1 1N4148 diode

1 2N3904 and 1 2N3906 transistor

2 10-kΩ, 2 5.6-kΩ, 1 1-kΩ, and 1 10-Ω resistor

1 0.01-μF capacitor

Build the sawtooth generator in Figures 6.9 and 6.10. Use a charging capacitor of 0.01 μF. The charging current is about 1 mA if the reference voltage is 5 V. Estimate the sawtooth frequency.

It is a good procedure to build and check each part of the circuit as you go. Build the constant current source first. Check the current level using a test resistor and a voltmeter. Add the clamping transistor so that it clamps the test resistor. Connect the base resistor of the clamping transistor to the common or to the positive supply voltage to turn the clamp off and on. Verify that the voltage across the capacitor is near 0 when the clamp is turned on.

Now complete the circuit. Observe the sawtooth waveform on the oscilloscope. Reduce the input resistor and observe that the frequency increases. Add to the capacitor and observe that the frequency decreases. Use the oscilloscope to measure the time it takes to discharge the capacitor. Estimate the current level during discharge.

Timing Circuit

The integrator circuit can be used to provide a time delay. The circuit in Figure 6.11 has a short across the charging capacitor.

When the switch opens, the voltage across the capacitor begins to ramp up. When this voltage reaches the comparator reference voltage

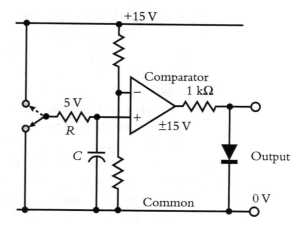

When the switch is in the up position, the capacitor begins to charge. When it reaches 5 V the comparator changes state and goes to +15 V. The comparator should have a high input impedance.

If the comparator is an IC amplifier operating from plus and minus power supplies, then the ouput diode keeps the output voltage from going negative.

Figure 6.11 A time-delay circuit

level, the comparator output goes positive and can operate an LED, a relay, or another transistor. The integrator resistor and capacitor control the delay time. A time delay is sometimes required for power-up, to limit access time, operate a temporary light, or control exposure time in photography. With the right IC amplifier, if the capacitor leakage current is low and the capacitance is high enough in value, the time delay can extend to several minutes.

Trigger Circuit

When a slowly changing signal hits a critical level, it is often desirable to make a positive decision to do something. The situation might be the overheating of a bearing. A trigger circuit responds to the critical temperature level and transitions from 0 to the plus power supply voltage. This voltage can then operate a relay to sound an alarm. The output will not transition back to 0 unless there is a significant reduction in the input signal. For example, in Figure 6.12 the negative input voltage is set to 5 V.

If the signal voltage reaches 5 V, the trigger circuit output goes from 0 to 15 V. If the signal falls below 3 V, the trigger circuit output transitions back to 0. The gap between 5 V and 3 V is called *hysteresis*. This circuit is known as a *Schmidt trigger*.

After the threshold is reached, the positive feedback from the output increases the input signal to 7 V. The circuit cannot reset itself until the input voltage drops back to 3 V. This hysteresis stops the trigger circuit from oscillating around the transition point. If there is oscillation the trigger voltage would be unusable.

A capacitor can be used to maintain hysteresis for a limited time. When a very short pulse triggers the circuit, the output voltage is held positive by hysteresis. After the capacitor charges, the hysteresis is removed

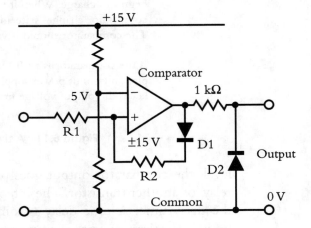

When the input voltage is less than 5 V, the output of the comparator is negative. Diode D2 limits the output to −0.6 V. Diode D2 is not conducting. When the input exceeds 5 V, the comparator changes state and goes positive. Diode D1 conducts and adds a positive signal to the input. This effect is called hysteresis. The resistor values can be selected so that the input must return to 2 or 3 V before the comparator resets. The circuit can also be designed so that R2 must be opened to reset the circuit. This condition occurs when R2 is a low value.

Figure 6.12 A trigger circuit to sound an alarm

and the trigger circuit returns to its initial state. The pulse width is controlled by the size of the capacitor. This type of circuit is called a *one-shot*.

The Multivibrator

A class of circuit can be made from IC amplifiers or transistors that have two operating states. As an example, an output voltage could be stable at 0 V and +15 V. These two voltages are often near the limits of the power supply. Multivibrator circuits that change state in response to a pulse or an input signal are called *bistable multivibrators* or "flip-flops." Circuits that have voltages that constantly transition back and forth between two end states are known as *astable multivibrators*. They are a class of oscillator.

Circuits that have two stable states are used extensively in digital logic. It is worth noting that digital ICs are made from analog circuits that include gain, output drivers, comparator clamps, zeners, and diodes. In the early days of digital electronics, all of the logic functions were built from discreet transistor circuits. These circuits are still useful, and it is helpful to know how they operate.

In the following sections we will discuss two circuits, the bistable and the astable multivibrator, which are analog circuits used in digital design. A multivibrator circuit can be recognized by its structure. It usually consists of two gain elements that are cross-coupled—gain element A drives gain element B and gain element B drives gain element A. The coupling is such that one of the gain elements conducts while the other is turned off. In the bistable multivibrator an input pulse will cause the gain elements to change roles. In the astable multivibrator the roles change automatically. After a transition a capacitor discharges, allowing the next transition.

The Bistable Multivibrator

Two cross-coupled transistors can form a bistable multivibrator. This circuit is shown in Figures 6.13 and 6.14.

Assume transistor Q1 is turned on and its collector voltage is low. This collector voltage couples to the base of Q2 through a resistor divider and keeps Q2 turned off. The high collector voltage of Q2 in turn keeps Q1 turned on. (We could also have started by assuming that transistor Q1 is turned off; in this case the roles of the two transistors would be reversed.)

A positive pulse at the input terminals couples to both bases. The

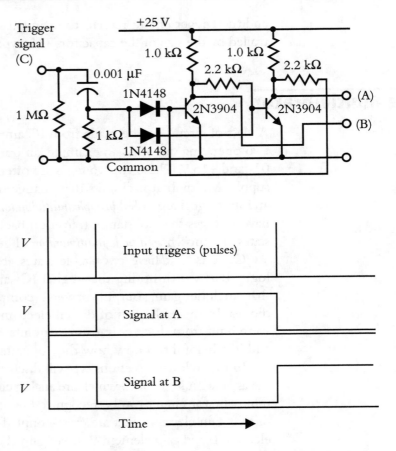

Touch point C to the power supply to generate a trigger.

Figure 6.13 A bistable multivibrator

pulse cannot add to the current in Q1, because the transistor is already fully conducting. The pulse can, however, cause Q2 to conduct. The collector voltage of Q2 drops. The voltage divider to the base of Q1 couples this voltage drop and reduces the current flow in Q1. The rise in the collector voltage of Q1 further increases the base voltage of Q2. The end result is that transistor Q1 is turned off and transistor Q2 is turned on. From symmetry, a second input pulse causes Q1 to conduct and Q2 to be turned off. In this way the circuit can be made to flip-flop back and forth.

IC Applications   203

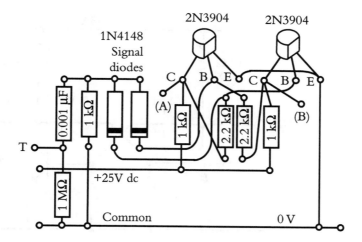

Touch point T to +25 V to create a trigger for the multivibrator.

Figure 6.14 The construction of the circuit in Figure 6.13

LEARNING CIRCUIT 42

Observing the Operation of a Bistable Multivibrator

You will need (in addition to your circuit board and measuring equipment):

- 2 2N3904 transistors
- 2 1N4148 diodes
- 1 0.001-μF capacitor
- 3 1-kΩ, 2 2.2-kΩ, and 1 1-MΩ resistor

1. Construct the bistable multivibrator in Figures 6.13 and 6.14.
2. Measure the voltage on the two collectors and determine which transistor is turned on.
3. To change the state of the mutivibrator, touch point C to the plus 15-V. This generates a pulse to change the state of the multivibrator.
4. Touch point C to the power supply a second time. The circuit should again change state. Verify this by observing the collector voltages.

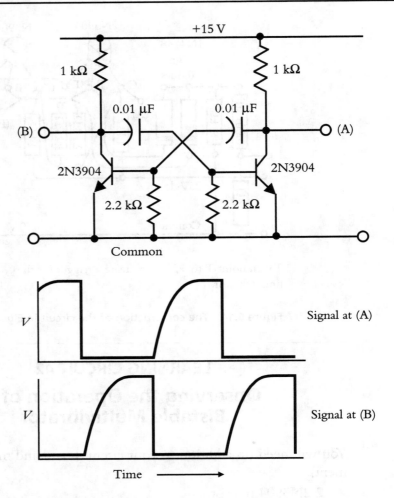

The time of one half-cycle is proportional to the RC time constant where R is the resistance from the base to common.

Figure 6.15 An astable multivibrator

The Astable Multivibrator

An astable multivibrator can be built from two IC amplifiers or two logic functions. The circuit shown in Figure 6.15 may not work all of the time, but it illustrates the principle. A breakdown occurs if both transistors happen to conduct equally at power-up. If this happens, just restart the power supply.

The moment Q1 starts to conduct, its collector voltage falls to near zero. The base voltage on Q2 also falls to a negative value. This is because C1 and R1 make a high-pass filter and the leading edge comes straight through this filter. As capacitor C1 charges through R1, the base of Q2 rises in voltage. When the base of Q2 is positive with respect to the emitter of Q2, it starts to conduct. The collector voltage on Q2 starts to fall, and this turns off transistor Q1. This double action causes the state of the mutivibrator to switch. During the time C1 is charging, C2 is discharging, and during the time C2 is charging, C1 is discharging. The *RC* time constants determine the frequency of oscillation.

Crystal Oscillators

The operation of most digital circuits depends on an accurate clock signal. A clock signal is a square wave voltage that transitions between the limits of a power supply at a known frequency. The multivibrator circuit described previously can be used as a clock signal. The frequency is determined by resistor and capacitor values. However, the operating frequency is difficult to adjust using this approach.

A crystal is a thin layer of quartz. Quartz has the property that a voltage is developed between two surfaces when pressure is applied. If a voltage is placed across the conductive surfaces, a strain results. If the surfaces of the crystal are plated, then the voltage resulting from a strain can be easily sensed with a high-input impedance voltmeter. A wafer of quartz with conductive surfaces has all the characteristics of a capacitor.

The quartz crystal is a mechanical structure that can store potential energy. The potential energy is stored in stress. This potential energy can transfer to kinetic energy when the crystal moves. All the elements of an oscillator are present. The crystal mass stores potential and kinetic energy. If energy can be supplied per cycle, the mechanical vibration can be sustained. The result is a sinusoidal voltage across the plates of the crystal. The crystal oscillator frequency is very stable, and very little energy must be supplied per cycle to sustain an oscillation. A typical circuit is shown in Figure 6.16.

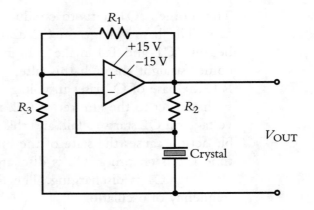

Figure 6.16 A crystal oscillator

The signal from a crystal oscillator is sinusoidal. To make a useful clock signal, the sine wave needs to be converted to a square wave. This can be done by using a comparator circuit at the oscillator output.

The 555 Timer

Several IC manufacturers produce a universal timing circuit. For example, National Semiconductor makes a component known as an LM555. This component can serve as a trigger, an astable or bistable multivibrator, and much more. The astable circuit can be used to change states after a given time. The technical literature shows how to add external resistors and capacitors to configure this device. Just like the IC amplifier, designers have considered many subtle factors and have produced a very useful and reliable component. The timer is inexpensive, compact, and very versatile.

Analog and Digital Representation

A voltage or a current level can represent parameters such as air pressure (sound), temperature, weight, speed, or time. These voltages or currents are called *analog signals.* Numbers that are stored digitally can represent these same parameters. Converting back and forth between analog and digital formats is an important part of electronics. These conversions occur so often that designers have created integrated circuits that perform the entire function.

An A/D converter is a circuit or component that converts analog (A) signals to digital (D) signals. A D/A converter is a circuit or component that converts digital signals to analog. Before we can discuss these converters, we must discuss how a voltage is represented digitally.

Counting digitally requires eliminating all the number symbols except 0 and 1. The first sixteen numbers are 0, 1, 10, 11, 100, 101, 110, 111, 1000, 1001, 1010, 1011, 1100, 1101, 1110, and 1111. Note that one of the numbers is a zero. Nothing is changed if we add leading zeros, so the sequence can also be written as 0000, 0001, 0010, 0011, 0100, 0101, 0110, 0111, 1000, etc. In a digital circuit, bits are represented by the states of transistors. A bit is a single logic state, a 0 or a 1. For example, if the transistor is turned on, it might stand for a 1, and if it is turned off, it might stand for 0.

Each digital position is called a *bit*. In this example, we have a four-bit number and we have counted from 0 to 15. If the counting had continued to include 8 bits, then we would be able to count to 255. Eight bits are called a *byte*. A string of 1s and 0s is called a *digital word*. A single bit could be 5 or 10 V, depending on the type of logic involved. The voltage itself has no meaning. The presence of a voltage represents a logic 1 and the absence of voltage represents a logic 0.

Consider a signal that varies from 0 to 6.4 V. We want to represent this voltage digitally. If we have a 4-bit digital word we can count to 16, so we can divide the 6.4 V into 16 parts. Each number represents one division, or 0.4 V. The number 0001 represents 0.4 V, and the number 0010 represents 0.8 V. The number 1110 represents 5.6 V, and the number 1111 represents 6 V. If one of the digital values is 0, then the largest decimal count is 15. The number 16 would be represented by all zeros and a carry bit. This is the same kind of carry we use in simple addition. Digitally the number 16 is represented by 0001 0000.

In order to represent voltages less than 6.4 V more accurately, more bits are required. If 8-bit words are used to cover the range 0 to 6.4 V, then an increment of 1 bit represents a change in voltage equal to 0.025 V.

The problem of converting a digital word into an analog voltage is one of converting a word like 1001 0011 into a voltage. Assume the least significant bit (the bit to the far right) represents 0.025 V. Each bit as we progress to the left represents a factor of two in voltage. The bits represent 0.025 V, 0.05 V, 1.0 V, 0.2 V, 0.4 V, 0.8 V, 1.6 V, and 3.2 V. The digital number 1001 0011 is the sum of 0.025 V, 0.05 V, 0.4 V, and 3.2 V. This sum is 3.675 V.

The R-2R Ladder

A very useful resistor network provides the basis for converting digital signals to analog signals. This network in Figure 6.17 is called an *R-2R ladder*. There are two values of resistor. The accuracy of the resistors is not as important as the ratio between them.

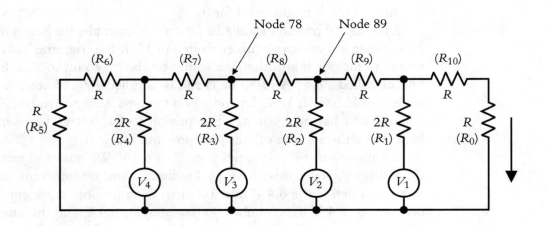

An R-2R ladder. If R_5 through R_{10} are 10 kΩ, then R_1 through R_4 are 20 kΩ.

If any node is opened, the resistance in either direction is always $2R$. For example, open node 78 and measure the resistance to common through R_7, and the resistance is 20 kΩ. The resistance through R_8 to common is also 20 kΩ.

If the current in R_0 for V_1 is 4 mA, then the current for V_2 is 2 mA and the current for V_3 is 1 mA. This assumes all the voltages are equal.

Figure 6.17 An R-2R ladder network

Typical resistor values might be 10 kΩ and 20 kΩ. Each of the voltage sources V_1, V_2, and V_3 circulate currents that flow in the ladder network. Each of these currents is independent, so we can look at them one at a time. The voltage source V_1 is shown in series with R_1, a 20-kΩ resistor. The current I that flows in this resistor splits equally between R_9 and R_{10}. The voltage source V_2 is in series with resistor R_2. The current in R_2 splits equally between R_8 and R_9. The current that flows in resistor R_9 splits equally between R_{10} and R_1. The current in R_0 is proportional to the current from V_1 and is one-half the current from V_2. This process can be extended to V_3. The current in R_0 is proportional to the current from V_1, one-half the current from V_2, and one-quarter the current from V_3.

Now assume all of the voltages are equal. This ladder circuit allows us to sum a group of currents from a fixed voltage source and attenuate these currents by factors of two. If the ladder has eight entry points, then the current that flows can be increased in steps where the largest step is 1 and the smallest step is 1/256. As an example, if the first current is 1 mA, the second current is ½ mA, and the third current is ¼ mA. Our problem is to control which voltages are to be connected. The maximum current in this arrangement is 1.996 mA. The R–2R ladder will only work if the voltage sources are set to 0 V when they are not used. A voltage source of 0 V is a short circuit.

The D/A Converter

Circuits that convert digital signals to analog signals are called D/A converters. The conversion can be accomplished using an R–2R network. The currents that flow in this network can be controlled by a set of transistor switches. The switches connect the 2R resistors to either a reference voltage or to 0 V (circuit common). The switches are controlled directly by the logic levels of the digital word. When the logic is a 1, the switch is turned on and a weighted current flows into the summing point. The amplifier responds to the sum of the network currents with a negative voltage that represents the digital word. A four-bit version of this D/A is shown in Figure 6.18.

The full-scale voltage for the D/A converter can be set by the reference voltage or by the gain of the operational amplifier. If a count of 15 equals 10 V, then each count represents 0.666 V.

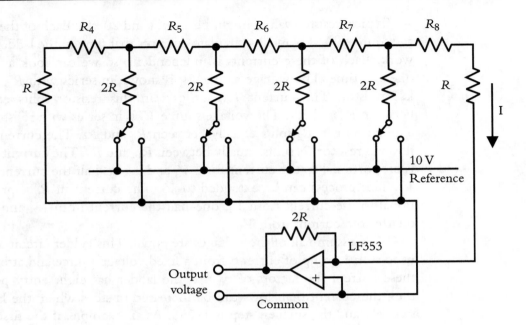

The IC amplifier operates from standard power supplies. A reference voltage is used in the R-2R ladder so that known currents flow to the summing point of the IC amplifier. Remember, the minus input to this amplifier is a virtual ground, and it does not move more than a few microvolts.

The switches shown are symbolic. Actual switches might be MOSFET logic that operate from external logic signals.

Most D/A converters contain the R-2R ladder, the IC amplifier, the reference voltage, and the logic switches.

Figure 6.18 A four-bit D/A converter

A/D Converters

There are several techniques available for converting an analog signal into a digital word. Accurate conversions at a rate exceeding 200,000 samples per second are practical. At these high speeds there is little time available to take a series of logical steps. This makes a high-speed A/D converter a very sophisticated circuit.

An analog signal to be converted to a digital word must first be sampled. The digital measure is made of the sample, not of the changing data. A *sample-and-hold* circuit provides the digital circuit a steady signal for conversion. The sampling circuit might consist of a switch, a capacitor, and a buffer amplifier that measures the voltage on the capacitor.

A very simple way to determine the digital value for an analog signal is to advance a digital counter through its set of values. A digital counter is a circuit that stores a set of bits in a circuit called a *register*. An external pulse advances the binary number stored in the register. The bits in the counter are converted to an analog signal (D/A converter) by using an R-2R network and an operational amplifier. This D/A converter output is compared with the sampled voltage. When a comparison is reached, the counter is stopped and the stored digital count is the correct digital word. There is always a round-off error, as there are only so many values available, depending on the number of bits.

Faster A/D Converters

Flash A/D converters are very fast. A sample-and-hold signal is applied in parallel to a number of comparators. It takes 15 comparators for a 4-bit A/D converter and 255 comparators for an 8-bit converter. There must be a comparator for every possible digital value. For example, if the resolution is 0.025 V and a voltage of 3.0 V is sampled, 120 comparators show a 1 and 135 comparators show a 0. This grouping of comparator signals is logically converted to an 8-bit word. This word is the desired digital value.

Another type of A/D converter is called a *pipeline flash converter*. The sampled signal is applied in parallel to a group of comparators. This results in an approximate measure of the signal. The output of these comparators is connected to a D/A converter. The output of this D/A converter is subtracted from the initial signal. The difference signal is amplified and again sent to the comparators. The output of the comparators is again connected to the D/A converter. The output of the D/A is again subtracted from the input. This process is repeated until all of the bits in the A/D converter have been determined. This pipeline approach takes more time than the true flash system, but it provides for a great deal of accuracy.

Successive approximation is another technique that is often used. The signal is sampled as before. The most significant bit in a digital register is placed in a D/A converter and the comparator is monitored. If the comparator indicates that the analog signal is too high, the bit is removed and the second most significant bit is tried. If the comparator says the compare signal is too small, this bit is left in place. This process is repeated for each bit in the digital word until the least significant bit is set.

The Wheatstone Bridge

Four resistors arranged in a square form a *Wheatstone bridge*. This configuration of resistors is used to make many measurements. The scales that weigh a loaded truck or the scales in the post office that weigh letters and packages both make use of this type of bridge. This bridge configuration is also used to measure strain, stress, and shear in mechanical systems. Other applications include vibration and temperature measurements. A typical bridge circuit is shown in Figure 6.19.

A dc voltage is placed across one diagonal of the bridge. This voltage is called the *bridge excitation voltage*. The difference voltage measured across the other diagonal is the signal of interest. If all four resistors are equal, this difference signal is 0 V. If one of the resistors changes value, a differential signal voltage results. Because the resistance from each signal connection to common is nearly the same, this signal is a balanced signal. If the excitation voltage is 10 V, the voltages at the two signal leads are 5 V. This average voltage is a common-mode voltage and must be rejected by any subsequent amplification. Only the difference signal is of interest.

One, two, or four arms of the bridge can be active. An active arm changes resistance when the parameter of interest changes. If one arm is active, then the other three arms are usually passive resistors. The resistors that make up the passive arms must not change value when the temperature changes.

For stress and strain measurements, the active arms are made from a long strip of thin resistance wire that is fan-folded and bonded to a plastic carrier. This active arm is called a *strain gauge*. It is oriented and bonded to a structure so that if the structure changes dimension, the gauge wire is stretched or compressed. The change in dimension changes the resistance of the strain gauge. If this gauge element is one

IC Applications   213

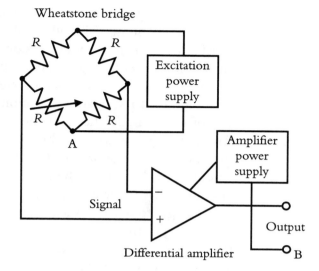

Point A is usually grounded at the item being measured. Point B is usually grounded by the recording equipment. The bridge element with the arrow is the active arm. The other arms may be located in the differential amplifier. The excitation power supply is often a separate circuit and not a part of the amplifier power supply. One half of the excitation voltage is a common-mode signal for the differential amplifier.

Figure 6.19   A typical Wheatstone bridge circuit

arm of a Wheatstone bridge, a difference signal is generated across the signal terminals of the bridge. Strain gauge elements can have a resistance in the range 50 to 300 Ω. Under stress, these resistors might change 1 or 2%.

The signal voltage from a bridge is proportional to the bridge excitation voltage and the percentage change in each resistance. The equation for the output voltage difference for one active arm is

$$V_{\text{Sig}} = (\Delta R/4R) \times V_E \tag{6.1}$$

where ΔR equals the change in resistance, V_{Sig} is the bridge signal, and V_E is the excitation voltage. The excitation voltage is a separate and well-regulated power supply.

A strain-gage element is calibrated in terms of resistance change per micro-inch of strain. This allows the mechanical engineer to equate a signal voltage to an actual dimensional change in a structure. Mechanical engineers select the points on a structure that should be monitored. As an example, two strain gauge elements can be placed in a bridge configuration to separate elongation from bending. In bending, one side of a beam stretches and the other side compresses. If gages are placed on opposite arms of a bridge, the bending changes the resistances but leaves the bridge balanced. If these same gauges are placed on adjacent arms of the bridge, the bending signal is reinforced and the elongation signal is canceled. In another application, it is possible to mount strain gauges to separate shear from strain.

The signal that is developed across a bridge is greatest when all four arms are active. Two gauges must increase in resistance and two must decrease in resistance to produce an optimum signal. If all gages increase in resistance by the same amount, the gauge remains balanced.

A strain gauge bridge can be used to measure large mechanical forces. An example might be a scale used to weigh a truck. A weighing platform is mounted on four support points called *load cells.* Four cells are required because the truck may not be centered on the platform. Each load cell is a full-strain gauge bridge. The signal from each load cell is amplified and summed at a central point. The total signal is a measure of the truck weight.

A strain gauge bridge located on a diaphragm can be used to measure pressure. As the pressure builds, the diaphragm expands, changing the balance in the bridge.

A piezoelectric resistor changes resistance depending on pressure. The pressure can result from the forces of acceleration. A group of these pressure-sensitive resistors in a bridge configuration can be used to measure vibration in large structures such as aircraft wings, buildings, and suspension bridges.

In some testing it is desirable to know the distance between two points. In automobile testing it might be desirable to measure the distance from an axle to the body of the car. A telescoping tube with a slider on a long resistor can be used to form two resistances in a bridge. When the slider is in the middle, the resistors are equal and the bridge is balanced. As the parts move, the slider moves, unbalancing the bridge. The resulting signal indicates the spacing between the objects.

A differential amplifier must be used to amplify the signals from a

Wheatstone bridge. When the gain that is required is less than 20, then an amplifier similar to the one shown in Figure 5.15 can be used. For higher gains, a variety of instrumentation-type IC amplifiers are available on the market. Technical notes are available that show the user how to balance the bridge and provide needed input protection.

SELF-TEST

1. The current source in Figure 6.2 has a 200-Ω sensing resistor. The input voltage is 3 V at 1 kHz. What is the output current? What is the output voltage if the load resistor is 500 Ω?

2. In problem 1, what is the output voltage if the load resistor is 400 Ω?

3. A current source in Figure 6.2 has a 400-Ω sensing resistor. The input voltage is −4 V at 3 kHz. What is the output current? What is the output voltage if the load resistor is 400 Ω shunted by 0.1 μF?

4. An integrator circuit uses a 0.1-μF capacitor and a 1-MΩ resistor. A 2-V dc input signal is applied. How long will it take for the voltage to reach 10 V?

5. A 20-Hz 10-V square wave signal is placed into the integrator of problem 4. What is the output waveform? What is the peak-to-peak output voltage?

6. A 350-Ω bridge is used to measure strain. The excitation level is 10 V. When the gage changes resistance by 1.5%, what is the differential signal?

7. In problem 6, how much gain is required to produce a 10-V signal?

8. In problem 6, if the strain measure must be 0.1% accurate, how much offset voltage is permitted?

9. A strain gage bridge uses 400-Ω resistors. The excitation is 5 V. Diagonal resistors are active. The first resistor increases 1% and the second resistor decreases by 1%. What is the bridge signal? *Hint:* Assume the signals are additive.

10. An 8-bit digital word represents voltages from −10V to +10 V. What is the digital value for 10 V?

11. In problem 10, if all 0s represent −10 V, what does the digital value 1000 0000 represent?

12. In problem 10, what is the voltage difference for the least significant bit? *Hint:* The count maximum is $2^8 = 256$.

13. Can the first bit be used to tell the sign of the voltage?

14. Can you find a way to use a digital representation of a voltage and separate the sign of the voltage from its magnitude of the voltage?

ANSWERS

1. 15 mA. For 500 Ω the voltage is 7.5 V.

2. 6 V.

3. The current is 10 mA. The impedance at 3 kHz is 267 Ω. The voltage is 2.7 V.

4. $Q = CV = 0.1 \times 10^{-6} \times 10 = 1$ μC. 1 V flowing in 1 MΩ for 1 second is 1 microcoulomb. 2 V applied for ½ second is also a charge of 1 μC.

5. The waveform is triangular. 10 V for 1 second is 10 μC. A half-cycle at 20 Hz is 25 ms. For 25 ms the charge is 0.25 μC. The voltage on 0.1 μF is $q/C = 0.25/0.1 = 2.5$ V. This is the peak-to-peak voltage.

6. The voltage is one-quarter the percentage resistor change times the excitation, or 62½ mV.

7. The gain is 160.

8. The offset is 0.1% of 10 V, or 1 mV.

9. 0.01×5 V/2 = 0.025 V.

10. 0000 0000 with a carry bit.

11. 0 V.

12. The range of voltage is 20 V. One part in 256 is 0.078 V.

13. Yes. When the first bit is 0, a negative number is represented.

14. To obtain the magnitude of the negative number, swap 1s for 0s and add 1. This is called the 2s complement.

7 Circuit Construction, Radiation, and Interference

Objectives

In this chapter you will learn:

- how electric and magnetic fields affect circuits
- some techniques for laying out circuits to avoid noise and interference
- how transmission lines function
- how a coaxial cable carries power to an antenna

We have now covered many basic circuits and seen how they operate. Of course, we have only been able to touch upon our subject, and have not been able to go into details of constructing these circuits. If you continue your study of electronics you will come to appreciate that in many applications performance depends on the exact arrangement of components. In other words, the component geometry is an important part of the design.

Particularly in designs that require long cables or the interconnection of many pieces of hardware, problems can arise related to interference, accuracy, or stability. In this chapter we will touch upon a few of the problems. These topics are important in understanding electronics, but they are not easy to demonstrate. Learning Circuits are not practical on this level—I can hardly ask you to build an antenna or install a long power line. Nor are the kind of demonstration problems I have given you at the end of the previous chapters practical on this level. However, I have chosen to discuss these topics anyway, even though without Learning Circuits or practice problems, the discussion may appear somewhat abstract. Still, the advice you will find in this chapter is the result of a great deal of experience. I hope you will continue with your study of electronics and that this advice will be useful to you.

Circuits and Fields

The circuits we have studied can be analyzed using Ohm's law, plus an understanding of transistors and integrated circuits, the components that allow us to amplify, rectify, control, and monitor various electrical signals. However, in many situations, circuits include structures that are not under our control. The earth, for example, is a conductor that is shared by all users. It has resistance, but it is not a resistor. The conduit in a building or residence may be grounded (tied to earth) along with our electronics. The interconnection of equipment grounds is not a part of any schematic.

Space can support radiation, but it is not a conductor, a capacitor, or an inductor. The behavior of an antenna cannot be described in terms of a circuit. Receiving a radio or television signal cannot be described in terms of components. Lightning is electrical, but there is no circuit to draw and analyze. Transporting signals in a long cable cannot be analyzed using simple circuits. These are all areas where circuit diagrams are simply not available. To understand them, you must have an understanding of electricity itself.

The next chapter discusses the basics of electricity. You will find that voltage is defined in terms of the *electric field*. Another field, the *magnetic field*, exists whenever there is current. Capacitance is the ability to store electric field energy, and inductance is the ability to store magnetic field energy. Whenever there is a voltage there is an electric field, and whenever there is a current there is a magnetic field. A changing electric field in a capacitor is a current, and a changing magnetic field in an inductor is a voltage. Both fields are present if there is any electrical activity. These fields exist inside components, but they also exist around all conductors involved in electrical activity.

In the circuits we have studied, the electric and magnetic fields outside of the components can be ignored. But when the frequencies extend beyond 1 or 2 MH, or if high accuracy is required, these fields cannot be ignored. When circuits extend between remote points, then fields cannot be ignored. Trying to draw circuits that describe the activity of fields is almost impossible. Getting a broad understanding of how fields behave is not difficult, but it can be elusive. Some of the problems encountered in building practical circuits involve electric fields. For this reason, it is important to appreciate how fields and circuits relate to each other.

Electrical Transport

All electrical activity involves both the electric and the magnetic field. The power that is sent to us by the utility company is carried not on the power wires, but in the fields between the wires. This may come as a surprise, but it is the only explanation that can be used to understand the many problems that do occur. In all circuits, fields carry all power and all signals. The conductors are guides that direct the path of signal or energy flow. It is difficult to use this viewpoint to design circuits, yet if we do not embrace it on some level, many interference processes will stay a mystery.

Electric fields store energy. Nature is always looking for ways to let energy flow to a lower state. An analogy with water may be helpful here. Water in a water tower stores energy of position, and will flow down to earth if there is a path. If the path involves turning a turbine, then the stored energy can do work. In the same way, stored electric field energy will follow conductors as a way to reach a lower energy state. The energy stored in a capacitor will heat a resistor if the resistor shunts the capacitor. A television signal is a field traveling in space. A television antenna provides a path so that some of this energy will flow to the receiver circuit. The field takes this path because it is flowing to a lower energy state.

The space around us is filled with fields. These fields include utility power, radio stations, television, radar, cell phone signals, telephones, and ham radios. Lightning and static discharge generate fields. As these fields travel in space, they can couple energy to conductor pairs and follow these paths into circuits. This is nature running downhill. A circuit operates on this same principle. Energy from the power supply is running downhill through the components to do our bidding. There are power leads and signal cables connecting to our circuits. Nature will couple field energy to these conductors. The coupled energy goes in both directions on every conductor pair, and some of this energy can enter our circuits. A power cable can carry power into a circuit, and at the same time it can transport interference in or out of a piece of hardware. The interference field can be transported between the power conductors or between the power conductors and earth.

In power transport from the generator, the neutral is connected to earth at many points. The majority of the power is carried in fields between the "hot" conductors. Some small percentage of the power is transported in the field between the power conductors and the earth. This means there is earth current involved in power transport. If there are buried conductors in the earth such as fences or gas lines, the current will tend to concentrate in these conductors. This means that power fields exist between the "hot" conductors and earth. This is the reason there is a concentration of power-related fields in all areas that use utility power. This explains why there will always be current flowing in the structural steel in buildings or in the conduit associated with residential construction. These currents have fields that extend into most electronic hardware. They originate from grounding or earth connections, so more earth connections provides very little relief.

If power could be transmitted coaxially (see the following section on transmission lines), the fields would be confined to the inside of a cable and could not get out. But this is not economical. Besides, there are always ways to solve power-related problems without resorting to coaxial power transport.

The Electric Shield

The simplest shield is a metal enclosure. The enclosure can take on many shapes; often it is a woven braid that extends over wires in a cable. The shield material can be aluminum, steel, or a conductive paint. If the shield surrounds a circuit, the electric fields associated with voltages in the circuit will be confined to the enclosure. Electric fields that are external to the shield will stay outside. Inside the enclosure, the effect of the electric fields can be equated to capacitances from the circuit to the shield surfaces. These capacitances function just like circuit components, allowing current to flow. The shield is usually connected to the circuit common to avoid unwanted coupling and capacitive feedback. If the shield is not connected and left floating, it functions as a capacitive divider and external fields can add signal to the circuit.

In an ideal shield enclosure, the fields outside the box are reflected and do not enter the enclosure. Of course, the ideal perfect enclosure rarely exists. Fields enter via the power leads and cross through the transformer coils. Fields enter on input and output leads. At high frequencies the fields can enter through seams and holes. Shielding that is effective at 60 Hz against the electric field may be ineffective against a magnetic field.

In many circuits it is only necessary to shield the input signal right up to the input of the circuit. This is electric field shielding. The rest of the circuit is often relatively insensitive to coupling. If the input stage adds gain and has a low output impedance, then coupling to the second stage can be ignored. It is often desirable to add a small RC filter right at the input terminals as added protection. Typical values might be a series RC circuit consisting of a 100-Ω resistor and a 100-pF capacitor.

The metal enclosure that surrounds a circuit is often connected to safety ground. In stand-alone hardware this green wire connection is required by the electrical code as protection against electrical shock. When a circuit enclosure is left ungrounded, it assumes a potential determined by the transformer and its many internal capacitances. This often causes the circuit to be very noisy. When an external ground is added, the noise appears to go away. This reduction in noise often leads to the conclusion that circuits perform best when they are well grounded. This in turn leads to elaborate grounding schemes for facilities. Unfortunately, grounding does not eliminate the fields that are the source of interference. It makes more sense to solve the problem for each signal than to try to find a global solution in terms of grounding.

Integrated circuits have the advantage of being very compact. This means that the component itself is relatively immune to external fields. This places the coupling burden on supporting components and on the cables that carry signals in and out of a circuit.

Common-Impedance Coupling

The power supply circuits that we have considered have diode rectifiers and electrolytic capacitors. It is not uncommon for currents in the electrolytic capacitors to be several amperes. This current can cause a voltage drop in the wiring. As a result, the common side of the capacitor may not be at the same potential as the centertap of the transformer. If a conductor has a resistance of 10 mΩ, a peak current of 2 A will develop a peak signal of 22 mV. If this signal is added to an input signal and amplified, the resulting hum can be very objectionable. Adding more capacitance will not cure the problem.

The power supply should be built so that the leads that carry the filter current are not a part of any signal circuit. Figure 7.1 shows a proper and an improper connection to a power supply.

Once an objectionable signal is mixed with a desired signal, it cannot simply be removed.

The common lead in a circuit is called the *reference conductor*. This conductor is often a trace with a very limited cross-section. A conductor is a reference conductor only if there is no significant voltage drop along the conductor length. In a circuit with a power supply, the reference point must be taken after the filter capacitors and after the regulators.

The wiring resistances between components can influence the performance of a circuit. It is usually possible to interconnect the components so that these resistances have little or no effect. If the resistance adds unwanted feedback or couples an unwanted signal, then the circuit operation may be unsatisfactory.

In a power amplifier output stage where there is feedback, the voltage drop in a common lead can actually change the feedback signal. In some cases this unwanted feedback can result in an instability. This problem can occur in output circuits that supply high current. The solution involves wiring the circuit so that the feedback signal does not sense the load current. Figure 7.2 shows a proper and an improper feedback connection in a power amplifier output.

Microphone cable consists of a center conductor and an outer braided conductor. This type of cable is a good example of the outer conductor serving as a shield and a signal conductor. If unwanted current flows in the shield it can generate a voltage that is added to the signal. In audio work very small amounts of interference can be disturbing.

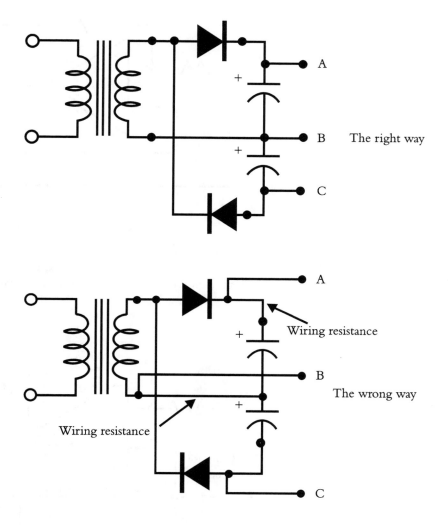

The right way is to connect loads after making the connections to the filter capacitors. The wiring of the power supply circuit has some resistance. The voltage drop in this resistance can appear in the output voltage and reduce the effect of the capacitors.

Figure 7.1 Power supply wiring

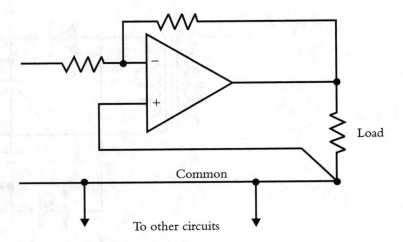

The correct way to connect the feedback elements of an amplifier. This is important if high-output current is involved.

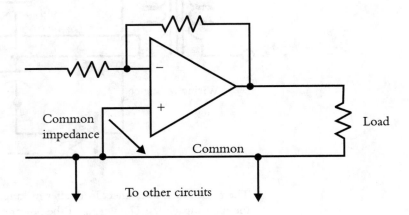

In this wiring arrangement the feedback senses the voltage drop in part of the common lead. The positive input to the IC is one of the feedback connections. This arrangement can sometimes be troublesome.

Figure 7.2 Feedback taken from different points in a circuit

Circuit Construction, Radiation, and Interference

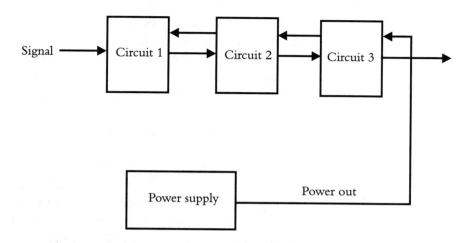

The general flow of power to various circuits is opposite to signal flow. This may not be important in many situations, but it is good design practice.

Figure 7.3 Connecting power to a circuit

The problem is apt to be severe if the signals of interest are small and require later amplification.

It is good practice in a circuit layout to connect the output stage directly to the power supply. This way the output stage current does not flow in the common connections associated with the input. Power should flow from the output of the circuit toward the input, and the signal should flow from the input of the circuit toward the output. This approach allows additional filtering for the input stages if it is needed. This technique is shown in Figure 7.3.

Separate circuits that demand extra current should be connected directly to the power supply. This technique reduces the risk of adding interference to sensitive circuits.

Star Connections

The conductors that are considered at 0 V can be numerous. The connections can include an input shield, an output shield, an input common, an output common, emitters, a transformer centertap, electrolytic capacitors, feedback connections, filter connections, a safety ground, and connections to the chassis. There are many ways to connect these grounded conductors together. One approach is to select a single point for grounding and bring all of these conductors to this common point. This is called a *star connection*. The idea is to eliminate any common resistance in the paths between various grounds. If there are no external grounds except the safety connection, then this idea can work. If the input shield is grounded at a remote point, current can flow in the input shield to the safety ground, and the star connection does not work. As indicated earlier, shield current can couple interference to an input signal if the shield is one of the signal conductors.

Single-point grounds are impractical in most circuit designs. The use of printed wiring precludes this kind of construction. The preferred approach is to arrange the connections so that interfering currents do not flow in sensitive circuits. Generating a star connection adds many loop areas to the signal path. These loop areas can couple to fields, and this can be another source of interference. We will examine this point in the next section.

The Signal Path

The signal of interest in an electronic circuit is a potential difference. The signal is always measured with respect to a reference conductor. As the signal progresses through the circuit, the reference conductor

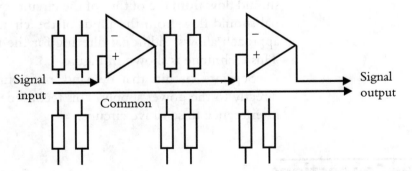

The best way to arrange signal flow is with a minimum loop area between the signal path and the signal common. This may not always be practical, but it should be a consideration, particularly in the first stages of gain or where small signals are handled.

Figure 7.4 The proper way to arrange a signal path

should follow the signal. In one sense the signal is funneled through the circuit in the area between the signal path and the reference conductor. The ideal design keeps this area minimized. The most critical areas involve the smallest signals. The theory is very simple: interference processes are proportional to area. Limiting the area in the signal path limits interference coupling. Circuit components that attach to the signal path should fan outward, leaving the signal loop area small. This arrangement is shown in Figure 7.4.

Transmission Lines

In the circuits we have considered, all power and signals are transported between two conductors. The signal is the potential difference between the input lead and the signal common. Power is also the potential difference between two conductors. As we have seen, power and signal can sometimes share the same conductors. Every pair of conductors is a potential transmission line. The power and signal voltages move so fast that for all practical purposes the time of transmission can be considered zero. But there are many situations where this viewpoint is inadequate. For this reason, a brief discussion of transmission lines is important.

The simplest transmission line, shown in Figure 7.5, consists of two parallel conductors and a load resistor. We are interested in what happens the moment the battery is connected to the line. We know that after a few microseconds the power supplied to the resistor is V^2/R. Before this can happen, field energy must travel along the transmission line to get from the battery to the resistor.

A fundamental property of nature makes it impossible to move energy in zero time. The presence of a field means there is energy stored in that field. The fields along the transmission line must start at 0. The

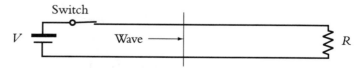

A transmission line consisting of two parallel conductors. When the switch closes, a wave is sent down the line.

Figure 7.5 A transmission line

moment there is a voltage placed across the transmission line there is some electric field. This field starts to move at about one-half the speed of light. In the first nanosecond the field covers the area between the conductors and extends 6 inches.

The capacitance per foot of line determines the charge that must be supplied to the conductors in that first nanosecond. Assume the capacitance is 10 pF per foot. If the voltage is 1 V, the charge q in the first nanosecond is 5×10^{-12} C. This means a current must flow equal to $q/t =$ 5 mA. In the second nanosecond the field has moved a total of 1 foot. The same current flows in that second nanosecond. The current that flows implies that a magnetic field is associated with the moving electric field. This combination of electric and magnetic field is called a *wave*. The wave front moves down the transmission line at about one-half foot per nanosecond. The current supplied by the battery is steady. This means that the battery reacts as if a resistor of $V/I = 200$ Ω were placed across its terminals. This value is called the *characteristic impedance* of the line.

If the line is 10 feet long, the wave reaches the terminating resistor in 20 ns. If the terminating resistor is 200 Ω, the battery continues to supply

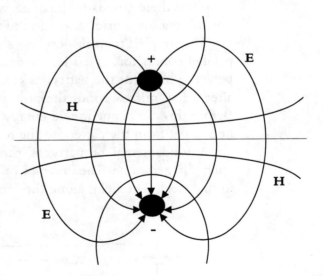

The electric and magnetic fields behind the wave front for a parallel conductor transmission line.

Figure 7.6 The electric and magnetic field lines around a transmission line

5 mA into the transmission line until the switch is opened. The energy that leaves the battery is converted from chemical energy to field energy before it is carried and dissipated in the resistor. It is important to realize that the fields carry the energy from the battery to the resistor. The wires direct where this energy is to flow. The wires do not carry the energy. The field pattern along the transmission line is shown in Figure 7.6.

The fields associated with the transmission line in Figure 7.6 extend out into space. Some of this field energy escapes and does not return to the circuit.

Coaxial Cable

Another kind of transmission line geometry is the coaxial cable. The two conductors involved in transmission are the center conductor and the outer cylinder. This geometry is shown in Figure 7.7.

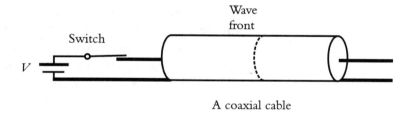

A coaxial cable

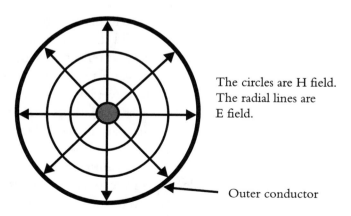

The circles are H field.
The radial lines are E field.

Outer conductor

Figure 7.7 A coaxial transmission line

In this transmission line the fields cannot leave the confines of the cable. Open wires are acceptable where the intent is to radiate the energy. The open parallel wires will also accept higher operating voltages than a coaxial cable.

The coaxial transmission line looks a great deal like a shielded cable. When it is used as a transmission line, the currents associated with the movement of the field flow on the inner surface of the sheath. This means that both ends of the sheath must be connected to allow this current to flow. If the cable is used for shielding against external electric fields, the outer covering may be connected (grounded) at one end.

Transmission Line Reflections

If the resistor in Figure 7.5 is removed, the transmission line is unterminated. When the switch connects the battery to the line, the wave propagates to the right at the same 6 inches per nanosecond. The energy moving in the line cannot be lost or stopped. When the wave reaches the open circuit, it must turn around and head back along the same path. There is no way for this energy to leave the confines of the transmission line.

Energy is now traveling in both directions at the same time. Because of the open circuit, the wave that is returned must cancel the current as it progresses back along the line. The voltage of the reflected wave adds to the voltage of the forward wave. An oscilloscope placed along the line would show double the battery voltage during the period of the first reflection. The battery continues to supply energy to the line until the reflected wave returns to the battery. At the battery, the returning voltage is incorrect and the wave is again reflected. This time the battery is shut off as the second reflection sends the first wave forward a second time.

If there were no losses in the transmission line, the energy would continue to travel back and forth forever. What really happens involves heating losses, radiation, and wave distortion. After a few round trips the reflections attenuate and the steady battery voltage appears along the entire line.

If the transmission line is terminated in a short circuit, the first reflected wave cancels the voltage at the end of the line. Wave energy is still returned to the source in the reflected wave. (Energy cannot be dissipated in an open or short circuit.) This time, when the returned wave reaches the voltage source, the battery must add more energy to the wave. The second reflection requires a battery current that is three times the initial current. Upon the fourth reflection, the current is multiplied by five. It is obvious that the current builds on each return reflection until a fuse blows or the conductors melt. This is the way current builds in a short circuit.

Sine Waves and the Transmission Line

When a sine wave voltage is applied to a transmission line that is terminated in its characteristic impedance, the voltage across the terminating resistor is the same sine wave, only delayed by the time of transit. If the transmission line is not terminated properly, the reflections return energy to the source. If the transit time is greater than one-twentieth of a cycle, the returning signal will modify the input impedance of the line. The input impedance then depends on frequency and the length of the transmission line. If the transmission line is terminated in an open circuit or a short circuit, the input impedance will be either a reactance, a short circuit, or an open circuit. If energy is to be transported effectively to the load, the characteristic impedance of the transmission line should equal the load resistance.

Common Transmission Lines

Considering transmission lines leads to an understanding of how signals and interference are transported. Nature makes no distinction between types of signals. We are the ones with a preference. We prefer clean information over noise; nature couldn't care less. The open conductors in a transmission line carry power or signals between two points. These same conductors can couple to any passing fields. This coupling may not be efficient, but it can add interference to the signals of interest. Coupled signals travel in both directions. Nature uses this coupling to conductor pairs to run downhill. It requires less energy to use the conductors than to stay in space.

There are many conductor pairs that nature can use. These conductors include the earth, shields, metal surfaces, conduits, oceans, and any open conductors, as well as our circuit conductors. Nature never reads labels, color codes, or directions. The available conductor pairs are not necessarily parallel, nor are they properly terminated. Some of them are brought directly into our hardware, and others add to the general ambient activity. The result is that coupled wave energy reflects and bounces around these conductors in a very complex and uncontrolled manner. It is much like the way light reflects and is absorbed by objects inside a room. The patterns are extremely complex.

Fortunately, most of the circuits we build cannot respond to these fields because the fields are changing so rapidly. Circuits are built to respond only to a particular portion of the coupled energy and ignore the rest. If we were building a cell phone, for instance, the circuit would be designed to respond only to the signal that carries the information the phone needs to function. This is the beauty of filters and resonant circuits. It allows the circuit to be very selective in how it responds.

Power lines that cross the space between buildings can couple to any radiated energy sharing the same space. The power conductors entering a circuit carry any and all types of signals. They can carry this signal right down the center of the conduit. These signals cannot do any work for us, but they can add to the interference that enters our circuits. The signals riding the power signals can be reflected or absorbed by line filters, provided the filters are properly installed.

Field Coupling

The most important tool in avoiding field coupling is to avoid a coupling area. Two conductors spaced 1 inch apart will couple to field energy proportional to the spacing and to the length of the run. To reduce the coupling, simply move the two conductors closer together. If the length of the run can be reduced, this is in the right direction.

In an earlier section we discussed how to route a signal and reference conductor in a circuit to reduce loop area. If the signal path and the reference conductor are kept close together, the loop area is automatically controlled. This practice will reduce coupling from the fields of nearby transformers or the fields from switching regulators as well as any fields that happen to enter the circuit.

If the fields are intense, then the next line of defense is to place the circuit inside a metal enclosure. A shield around a cable or around a circuit can be very effective. If the unwanted field enters the cable at either end, the shielding will be ineffective. Interference can use the space inside a cable for transmission just as easily as a desired signal can. If shielding is to be effective, all the points of entry must be considered. The problem is similar to the leaky boat. If some of the holes are plugged, the boat will still sink. All of the holes must be considered.

Here is a short story to illustrate the nature of the cable problem. A circuit consists of a sensor, a cable, and an amplifier. The connecting cable is shielded, and the shield is bonded to the metal boxes that contain the sensor and the amplifier, yet every time a nearby relay is operated, the circuit malfunctions. What was the design error? It turns out to be quite simple. A wire was placed inside the cable that connected the two metal boxes together. When this wire was removed, the circuits functioned properly.

But why did this wire cause a problem? Because the relay contact that opened created a field that used the cable and the earth as a transmission line. The current in this loop would normally stay on the outside of the shield used to cover the connecting cable. By adding a wire inside the cable, some of current in this loop used the wire. This created a field inside the cable that coupled to the sensor signal pair. When the relay opened, the sensor line sensed a signal that was coupled to the amplifier. By removing the wire, the interference field stayed outside the cable.

Radiation

Most of the circuits we have considered in this book operate at frequencies below 1 MHz. They transport energy between components and along a signal cable. These signals and their fields stay confined to the wiring. At 60 Hz most of the energy that leaves a generator reaches the expected load. If power were generated at 1 MHz and placed on the power grid, most of the energy would leave the confines of the conductors in the first 100 feet. Of course, this energy would disperse in all directions and would not be available to do the things we want to do with utility power. If we wanted to transport power between two conductors over any distance at 1 MHz, a coaxial cable would be needed.

When a transmission line is terminated in an open circuit, the energy that reaches the end of the line is reflected. If the center conductor extends beyond the shield, energy can leave the confines of the cable and it cannot get back in. This extended center conductor is called a half-dipole antenna.

The antenna radiates most effectively when the antenna length is a quarter-wavelength Energy leaves the antenna at the speed of light, which is 300 meters per microsecond. At 1 MHz a quarter-wavelength is 75 meters. This is the length of an antenna used to broadcast an AM radio signal. These large antennas are often seen as vertical structures on the outskirts of a city. A cell phone operating at 500 MHz would have an antenna that is 5.9 inches long. Such transmitters are now part of our everyday life. Antennas poke out from behind people's ears as they walk on the street or ride in their cars.

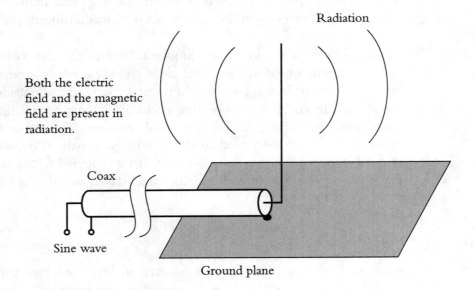

The half-dipole antenna. The sine wave is called a *carrier*. The power that is radiated is carried in the coaxial cable. The length of the antenna depends on the frequency. A typical antenna is one-quarter of a wavelength long. A quarter-wavelength at 1 MHz is 75 meters. At 500 MHz the length is 5.9 inches. This is the length of a cell phone antenna.

If the carrier amplitude is changed at the audio rate to send voice or music, the transmission is called amplitude modulator, or AM. If the frequency is changed, the transmission is called FM, or frequency modulation. The carrier by itself carries no information.

In television the voice is sent FM and the video pattern is sent AM.

The ground plane can be an auto body, a human being, an aircraft frame, or the earth.

Figure 7.8 The half-dipole antenna

The radiation from antennas is a very important part of electrical engineering. In most cases the person entering electronics for the first time is trying to avoid building a radiator. It is important to recognize how a radiator functions. Recognizing the effects of radiation comes from experience. It is hoped that the circuits that you will build do not radiate or couple to radiation unless this is your design objective.

The half-dipole antenna is not the only conductor geometry that radiates. The other basic geometry is a loop of wire. If the current loop is a quarter-wavelength in diameter, the loop will radiate energy very efficiently. The high current requirements make this a less desirable antenna design. This is another reason why loop areas in a signal path are undesirable. They can function as radiators.

Circuits that do not confine their fields will radiate. The radiators might not be efficient, but they can interfere with nearby pieces of electronics. The main reason the FCC requires radiation testing on electronic devices is to make sure they do not interfere with radios and television reception. The standards used in Europe go far beyond this requirement. Circuits will not radiate efficiently unless there is frequency content above 1 MHz. This can occur with certain types of switching regulators or with circuits that are oscillating. When there is arcing, there is usually radiation. Digital circuits with their high-speed logic can also radiate.

In Conclusion

This last chapter has touched upon many new areas, and includes topics that form the subject of many different books. If your interest is caught by this kind of discussion, you might enjoy two more of my books, *The Fields of Electronics* and *Grounding and Shielding, 4th Edition* (both published by John Wiley & Sons). They are written for electrical engineers, but you should now have enough experience with basic electronics to be able to read them if you wish.

If you've enjoyed building the Learning Circuits and found the discussions interesting, electronics could be a good career choice. However you decide to pursue your interest in electronics, I hope this book has given you a good start, and that you've had some fun along the way.

8 A Review of Basic Electrical Concepts

Introduction

To understand electronics, the study of electrical circuits in which voltages and currents perform useful tasks, you need some familiarity with electricity itself. This chapter is a review of electrical concepts, but if you are completely unfamiliar with them, the material here may not be sufficient. If so, I recommend that you read my companion volume to this book, *Basic Electricity*. This book is also part of the Self-Teaching Guides series, and will give you a step-by-step course in the subject, with problems in each chapter to help you become familiar with the material.

In addition to an understanding of basic electrical concepts, you also need some knowledge of algebra to understand this book. In any study involving electricity, concepts are often expressed in the form of algebraic equations. In other words, letters and symbols are used to represent components and variables used in electronics. I have also provided a brief review of simple algebra in appendix II. Again, if you are completely unfamiliar with algebra, the material in this appendix may not be sufficient. There are many books on algebra to help you get started, and many courses available on the high school and college level. In addition, some understanding of trigonometry will be helpful in sections of the book, but it is not required.

Electrons, Conductors, and Insulators

The story of electricity starts with the electrical forces that hold all atoms together. One of these forces is the electric force. This force exists between the electrons and protons in every atom. Molecules are formed when outer-shell electrons are shared between atoms. In metals, atoms are packed tightly together and the electrons are not shared. For this reason, the outer-shell electrons can easily move between atoms. These are known as *free electrons*. Materials with free electrons are called *conductors*. The most common conductors used in electronics are copper, aluminum, and iron. Materials whose electrons cannot move freely are called *insulators*.

Electrons have two qualities: an electrical charge and an electric field. The electric field is a region of influence (force) that can move other electrons. Within an atom the fields from the electrons are exactly balanced by the fields from the positive charge of the protons in the nucleus. When there are extra electrons, their electric fields combine to form a larger field. This field extends outside the atoms into the space around the object.

A group of excess electrons is also called a *negative charge*. The number of electrons that participate in electrical activity is a very minute fraction of the electrons that are available. The electric force is a very potent force. If 1% of the electrons in a human were to be active, the force they represent would be sufficient to move Earth out of orbit. This implies that in all electrical activity a very small fraction of the available electrons are involved.

When two insulators are rubbed together, charges can rub off one surface and deposit on the other. For example, rubbing a silk cloth over a plastic rod will cause some electrons to move from the cloth to the rod. Once electrons are added to the rod, they can be transferred to other objects through a contact.

If two conductors that hang from a string are touched, they will repel each other. This repelling force is attributed to the electric fields that surround these added electrons. This repelling force field acts at a distance on every electron in the vicinity. Free electrons in nearby conductors will move as a result of this force.

The absence of electrons also has an electric field. The absence of electrons is equivalent to a *positive charge*. Two objects with a positive charge will repel each other. Objects with a negative charge will attract objects with a positive charge.

If a negatively charged object is brought near a conductor, the free electrons on the conductor will be influenced by the external electric field and move away from their points of equilibrium. These electrons will move until there is a balance of forces between the displaced electrons and the external field. The electrons that move away from a part of the surface leave behind a field that for all practical purposes behaves like it is associated with positive charges. These pseudo–positive charges have the same mobility as actual electrons. It is possible to have areas of positive and negative charge on the same conductor.

In a static situation, extra electrons on a conductor distribute themselves on the surface of the conductor. If there were a field inside the conductor, then free electrons would be in motion, and this is not a static condition. Similarly, in a static situation there can be no electric field parallel to the surface of a conductor as this would also cause electrons to move on the surface. The number of electrons that are involved in any electrical activity is a fraction of those available.

Charge and the Electric Field

The unit of charge is the *coulomb*, abbreviated C. The letter q or Q is often used to represent a given charge. One coulomb of negative charge is 6.28×10^{18} electrons. In most circuits the amount of charge involved is expressed in μC, which is 10^{-6} coulombs.

The field that surrounds a charged object is called an E or electric field. This field can exert a force only on another field. To measure the strength of a field, a second charged object must be used. The charge on the second object must be very small compared to the initial charge, or the field being measured will be modified. The field is measured by noting the force on this small test object. At every point in space the force has a direction. If arrows are used to represent the direction of the force, the stems of the arrows can be interconnected to form electric field lines. By convention, field lines start on positive charges and terminate on negative charges. These field lines can be used to map the character of the field between charged objects.

Work and Energy

Energy is stored in capacitors and inductors. This energy can do work by heating a resistor or turning a motor. Voltage is defined in terms of work. Before we can define voltage, we must define work.

The definition of work is "force times distance." The direction of the force must be along the path of motion. When a mass is lifted on Earth, the work is equal to the mass times distance. When 1 kilogram of water is lifted 1 meter, the work done on the water is equal to 1 kilogram-meter. This work is stored as *potential energy* or energy of position. Potential energy has the same units as work. The energy stored by the water can be used to do work by releasing the water and letting it turn a water wheel.

A force acting on a mass can also accelerate that mass. An example is a car on rails. The work done on the car is again force times distance. In this case the motion of the car stores energy. Energy of motion is called *kinetic energy*. The units of kinetic energy are the same as the units of work. This energy of motion could be converted to energy of position by letting the car roll up an incline.

In electrical systems the electric field exerts a force on a charge. If a force moves this charge over a distance, work has been done on the charge, and potential energy is stored in the system. In electricity the unit of work is the *joule*. One joule is equal to 0.102 kilogram-meters (a force of 0.102 kilogram acting over a distance of 1 meter).

Voltage—A Field View

The definition of voltage involves moving electrical charges in an electrical field. Referring to Figure 8.1, assume a negative charge $-Q$ (electrons) is moved from the top conducting plate to the bottom conducting plate. These added electrons will spread out over the bottom surface. The absence of electrons (positive charge) will spread out over the top surface. This conductor geometry provides a near uniform electric field between the two conductors.

Consider a charge q that is much smaller than a charge Q. If the charge q is placed on a test object and the test object is placed in an elec-

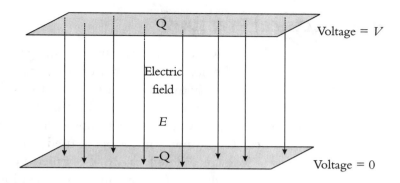

Figure 8.1 The electric field between two conductors

tric field, the force in the object is the intensity of the electric field strength times the test charge, or

$$f = E \times q \tag{8.1}$$

In Figure 8.1, the E field is uniform and the force on the test object is constant. The work required to move the test object between the two conductors is $f \times d$ where d is the spacing. The work W is therefore

$$W = E \times q \times d \tag{8.2}$$

where E is the field strength, q is in coulombs, and d is in meters. By definition, the ratio of work to charge is the voltage V. In equation form:

$$V = W/q = E \times d \text{ and } W = q \times V \tag{8.3}$$

where V is the voltage difference between the conductors, W is in joules, and q is in coulombs. Since $E \times d$ must have units of volts, the E field must have units of volts per meter. A voltage difference is a measure of the work required to move a charge between two points.

In most circuits, the E field is not a part of the calculation. However, you should realize that charges respond to the E field and not to voltage.

Voltage (Electrical Pressure)

Voltage is electrical pressure. A voltage difference implies an electric field. The field actually provides the force that moves electrons. It is customary to say that voltage causes electrons to flow in the conductors and components of a circuit. Any point in a circuit can be called 0 V. This simply means that voltages only exist as differences. If a point is at 10 V, then some other point is being used as the 0 or reference point.

Voltage differences are also called potential differences. The term *electromotive force (emf)* is sometimes used instead of voltage. Voltage differences can exist in space. Voltage is not limited to points on conductors. Of course, the measurement is hard to make, but it is a part of the definition of voltage.

The definition of voltage involves an electric field and charges on the surface of the conductors. If there is a voltage difference, then a field and a charge distribution must be present on conductors. The two go together and cannot be separated.

The voltages in electronics can cover a wide range, and many units are used in order to cover this range. The millivolt or mV means 0.001 V. The microvolt or μV is 0.000001 V. The kilovolt or kV is 1,000 V. The megavolt or MV is 1,000,000 V.

Current

Current is the motion of electrons. To get electrons to move in a conductor, they must be in an electric field. The electric field exerts a force on the electrons that causes them to accelerate. This is comparable to a mass in a gravitational field. A free mass will accelerate until it hits the earth. An electron in an electric field, say, one inside a television tube, will accelerate until it strikes the phosphor on the screen.

An electric field will cause free electrons in a conductor to accelerate. They cannot accelerate far, however, as they will immediately collide with atoms. In a conductor, electrons can achieve an average velocity, but they cannot accelerate. The energy they gain in acceleration is immediately given up in agitating atoms. An increase in atomic motion is the same as a rise in temperature.

A steady flow of electrons is a *current*. A steady current in a conductor uses the entire cross-section of the conductor. This flow of charge is not limited to the surface. The unit of current is the *ampere*. One ampere is 1 coulomb of charge passing a given point in 1 second.

The equation for current is

$$I = q/t \qquad (8.4)$$

where I is in amperes, q is in coulombs, and t is time in seconds.

The letter A is often used as an abbreviation for current. A power breaker might be rated 10 A, which is read as "ten amperes." In a circuit the usual abbreviation for current is the letter I. A varying current is often given the letter symbol lowercase i. The units of current used most often in electronics are the milliampere or mA, which is 0.001 A, and the microampere or µA, which is 0.000001 A.

The Direction of Current Flow

Electrons are attracted to a positive potential. It would seem natural to say that the direction of current flow is the same as the direction of electron flow. Historically, however, the opposite convention was established, and it has persisted. In all circuit analysis, current is said to flow *from* a point of positive potential *to* a point of more negative potential. Electrons flow in the opposite direction, *to* points of higher positive potential.

The Resistor and Ohm's Law

A resistor is a component that limits the flow of free electrons. Resistors play a central role in all electronics. When a voltage is placed across the terminals of a resistor, the electric field in the resistor causes electrons to move. The average electron velocity will depend on the materials used to form the resistor.

There are many types of resistors, depending on application. The common carbon resistor is made from a mixture of powdered carbon and a nonconducting filler. The resistance is varied by changing the percentage of filler. This mixture, along with two connecting copper leads, is compressed into a small cylinder. A plastic housing with a resistance color code is then molded around the cylinder.

The unit of resistance is the ohm. When an electrical pressure (voltage) of 1 V is placed across a 1-ohm resistor, 1 A will flow. When an electrical pressure of 1 V is applied to a 2-ohm resistor, the current is ½ A.

The relationship between voltage, current, and resistance is *Ohm's law*. This relationship can be written in three different equations:

$$I = V/R \tag{8.5}$$

$$V = IR \tag{8.6}$$

$$R = V/I \tag{8.7}$$

where I is in amperes, V is in volts, and R is in ohms.

The common carbon resistor covers the range from 10 Ω to 22 MΩ, which is 22,000,000 Ω. The Greek capital letter omega (Ω) is the abbreviation for ohm. The other standard abbreviations are kΩ for 1,000 ohms, MΩ for megohm, and mΩ for thousandths of an ohm.

The circuit symbol for a resistor is:

Power

Power is the rate of doing work. In equation form this is

$$P = W/t \tag{8.8}$$

where P is power in watts, W is work in joules, and t is time in seconds. The work in moving a charge through a potential difference was given by $W = qV$. If both sides of this equation are divided by time, then W/t equals power P and q/t equals current I. This means that

$$P = VI \tag{8.9}$$

where P is in watts, V is in volts, and I is in amperes.

Power Dissipation in Resistors

The power dissipated in a resistor can be determined from Equation 8.9 if you substitute current or voltage from Ohm's law. Since $V = IR$ or $I = V/R$

$$P = I^2R \text{ or } P = V^2/R \tag{8.10}$$

Capacitors

A capacitor is a component that has the ability to store a charge that is proportional to the applied voltage. The parallel plates in Figure 8.1 represent a simple capacitor. Capacitance is the ratio of charge to voltage, or

$$C = Q/V \tag{8.11}$$

where C stands for capacitance in farads, Q is charge in coulombs, and V is voltage in volts.

The use of the letter C for both capacitance and charge can be confusing. Which meaning is intended can always be inferred from context.

The circuit symbol for a capacitor is:

$$\dashv\vdash$$

The farad is a very large unit, and it is common to use one-millionth of a farad, abbreviated µF. Even this unit is often too big. A smaller unit is the pF (picofarad), which is one-millionth of a µF. In exponential notation, 1 µF = 10^{-6} F and 1 pF = 10^{-12} F.

The voltage across the plates of a capacitor implies that there are charges on the plates and that there is an electric field between the plates. If the charge increases linearly, then the voltage also increases linearly. If both sides of Equation 8.11 are divided by time t, the term Q/t is current and the equation becomes

$$I = C \times V/t = C \times \text{volts/second} \tag{8.12}$$

This equation states that a steady current flowing into a capacitor results in an increasing voltage. If 1 A flows into a 1 F capacitor, the voltage rises at 1 V per second. If 1 µA flows into a 1 µF capacitor, the voltage rises at 1 V per second. When V/t changes, the current adjusts to the new value.

The Energy Stored in a Capacitor

The conductors shown in Figure 8.1 form the plates of a capacitor. To calculate the energy stored in this capacitor for a given voltage, we divide the stored charge Q into many smaller charges q. The plan is to move the charge across the space in small increments. When the charge on the plates is 0, it takes no work to move the first charge from the top plate to the bottom plate. The second charge q takes a small amount of work, as there is now an electric field. From Equation 8.3, the work required to move the last charge q is qV. The total work is the average work, or

$$W = 1/2 \times Q \times V \quad (8.13)$$

This equation can be related to the capacitance by substituting Q from Equation 8.11:

$$W = 1/2 CV^2 \quad (8.14)$$

Capacitors store energy in the electric field between the plates. This energy is not dissipated. It is stored as potential energy—that is, energy that can do work later. The ideal capacitor will hold stored energy until a path is provided for the charge to leave. A parallel resistor can provide this path. The stored energy will simply heat the resistor.

Practical Capacitors

The plates in Figure 8.1 would form a very small capacitor. Capacitance increases with larger surface area and reduced spacing between the plates. The capacitance also increases if the space between the plates is filled with an insulating dielectric rather than air. Typical dielectrics are glass, mica, Mylar, and polypropylene. The increase in capacitance brought about by using one of these materials is called the *dielectric constant*.

The required plate area is often obtained by rolling a sandwich of two metalized dielectrics into a tight cylinder. Capacitors larger than 1 µF are often made using a polarized dielectric. These dielectrics can only be used for one polarity of voltage, such as found in a power supply. These components are called *electrolytic capacitors*.

The Inductor and the Magnetic Field

An inductor is a circuit component that stores magnetic field energy. To understand an inductor, it is necessary to understand the magnetic field. When current flows in a conductor, there is a magnetic field around that conductor. This field can be demonstrated by noting the movement of a compass needle near the conductor. The magnetic field from a current is called an *H field*. When the conductor is wound around a cylinder (solenoid), the magnetic field is increased by the number of turns. When current flows, the magnetic field threads through the coils along the axis of the cylinder. Many inductors are manufactured by wrapping coils of wire over a small cylinder.

Magnetic flux is the magnetic field intensity times a cross-sectional area. It is represented by field lines that close on themselves. The total magnetic flux that is developed is proportional to the number of turns in the inductor, and to the current. In the solenoid, when the current increases, the flux that threads the turns increases. This increasing flux develops a voltage at the terminals of the solenoid that opposes the change in current flow. The opposing voltage is proportional to the rate of change of current and to the inductance of the solenoid. This effect is known as *Lenz's law*. In equation form:

$$V = L \times \text{(rate of change of current)} = I/t \qquad (8.15)$$

where V is in volts, current is in amperes, time is in seconds, and L is inductance in units of henries. This equation states that if the voltage is constant across an inductor, the current must increase at a fixed rate. This is directly analogous to the capacitor. A steady current flowing into a capacitor will result in a voltage rising at a fixed rate. If the voltage rises at a faster rate, then the current must increase.

A magnetic field can be described two ways. One way is the induction or B flux, and the second way is the H flux that is created by current flow. The changing induction flux induces the voltage stated in Lenz's law. For a given current, the induction flux can be increased by placing magnetic material in the flux path. This is exactly analogous to the increase in charge storage when a dielectric is added to a capacitor. The ability of a material to increase the induction or B flux is called *permeability*. Iron has a permeability that can exceed 10,000. This property of iron makes transformers and motors practical at 60 Hz. By having iron in the magnetic path, the current required to establish the induction flux can be held to practical limits.

The circuit symbol for an inductor is

Typical values of inductance are the millihenry, abbreviated mH, and the microhenry, abbreviated µH. 1 mH = 0.001H. 1 µH = 0.000001H.

The Energy Stored in an Inductor

The energy stored in an inductor can be calculated by considering a fixed voltage placed across the terminals. The energy for the first increment of time is $V \times i \times t_1$ (this is power times time). The energy for the last increment of time is $V \times I \times t_n$. The average current is $I/2$. The total energy is the sum of all the increments of energy, or $W = \frac{1}{2} V \times I \times t$. The steady voltage by Lenz's law is $V = L \times I/t$. Substituting this value yields

$$W = \frac{1}{2} LI^2 \qquad (8.16)$$

This energy is stored in the magnetic field. The energy is constant as long as the current is sustained. In practice, however, a sustained current can only occur in a superconductor where the resistance of the coil is 0. If the energy stored in a capacitor is considered potential energy, then the energy stored in an inductor by a current can be considered kinetic energy. The moving charges store the energy.

Transformers

A transformer is an electrical component that couples voltages by using a changing magnetic field. It consists of coils of wire wrapped around a magnetic path, usually made up of iron. A voltage on the first or primary coil creates a changing magnetic field in the iron. The flux from this magnetic field threads through a second coil. This changing flux induces the same voltage waveform on the second coil.

The voltage is proportional to the number of turns. If there is 1 V per 30 turns on the primary coil, there will be 1 V per 30 turns on the secondary coil. If there are one-half the total number of turns on the secondary, the voltage induced on the entire secondary will be halved.

Practical transformers have many limitations. A magnetic material can support only a limited amount of induction flux before it saturates. The turns of wire offer resistance. The current that creates the induction flux can affect the voltage waveforms. Power transformers will only function over a very limited frequency range.

The symbol for a transformer is

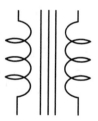

Semiconductor Materials

Electronics today is dominated by semiconductors. Components like transistors and diodes are made from semiconductor material. This same material is used to make integrated circuits or ICs, which may incorporate over 1,000,000 transistors. Integrated circuits provide many of the functions that at one time had to be built out of separate components.

The most common semiconductor material is silicon. Silicon is an element found in most of the rocks on Earth. Crystals of nearly pure silicon can be grown as ingots, which are then sliced into thin wafers. Thousands of components can be fabricated on one wafer. Fabrication consists of many steps that include vacuum deposition and etching using different optical masks.

A pure silicon crystal is an insulator. This is because the outer-shell electrons of the silicon atoms are shared to form the crystal. A phosphorous atom has one more electron than a silicon atom. If an atom of phosphorus is added to the silicon crystal, the extra electron becomes a free electron, and the silicon becomes a conductor. The addition of phosphorous atoms to silicon is called *doping*. The crystal is said to be doped with a donor atom. The doping levels can be as small as one part in 10 million. A crystal doped with phosphorous is called n-type silicon.

If boron is used as the dopant, then the material becomes p-type silicon. Boron has one less electron than silicon, so instead of providing a free electron, it provides a receptor site. The absence of an electron is called a *hole*. P-type silicon is also a conductor, as the holes behave very much like the pseudo–positive charges on a conductor.

If a sandwich of p and n silicon is created, the junction between the p and n material forms a *diode*. The diode is a common electronic component. Electrons in the n material can easily cross over into the p region if the E field is in that direction. In the opposite direction there are no free electrons in the p region to move into the n material, so in this direction the material behaves like an insulator. Conduction in one direction only is called *rectification*. The direction of easy current flow is called the *forward direction*. The direction of no current flow is called the *reverse direction*.

A sandwich of pnp or npn material forms a transistor. The outer two materials are called the *emitter* and the *collector*. The center material is called the *base*. If an electric field is impressed across the sandwich, the two diodes inhibit current flow in either direction.

In an npn transistor, the electrons flow in the forward direction through the base emitter diode, and then the sandwich acts like all n material. Electrons flow across from the emitter to the collector. This assumes there is a collector voltage. There is a multiplication effect by which a small amount of base current results in a significant amount of emitter to collector current. This multiplication of current is called *transistor action*. In a pnp transistor the exact same multiplication occurs, except that all the current directions are reversed. This action of transistors makes them very important components in electronic circuits.

Radian Measure of Angle

In electricity the measure of angle is the *radian*. Consider an arc on a circle equal in length to the radius r. Now form an angle by connecting radial lines to the ends of this arc. The angle formed this way is 1 radian. Since the circumference is $2\pi \times r$, there are 2π radians in 360 degrees. π is a constant equal to 3.1416. One radian is equal to 57.29 degrees.

The angles that are frequently used in electricity are 45°, 90°, 180°, and 360°. When measured in radians, these are $\pi/4$, $\pi/2$, π, and 2π radians.

Frequency

The unit of frequency used in electronics is the *hertz*. One hertz is equal to one cycle per second. The cycles can be of any recurring event. The abbreviation for hertz is Hz. The expression kHz is read kilohertz and means 1,000 cycles per second. The abbreviation MHz is read megahertz and the expression GHz is read gigahertz. One GHz is equal to 1,000 MHz.

Sine Waves

Sine wave voltages and currents play an important role in all electronics. Circuit analysis uses sine waves, as this is the only waveform that remains unchanged throughout most circuits. Sine waves are also used as carrier signals in most communications channels.

The sine function is one of the trigonometric functions. It is defined using a circle with a radius of 1. The radius of the circle forms an angle Θ with the horizontal axis. The sine of the angle is the height of the radius tip above the horizontal axis. As the tip rotates counterclockwise, the height h varies in a sinusoidal manner. Mathematically this is stated as

$$h = \sin \Theta \qquad (8.17)$$

The sine wave and a rotating radius are shown in Figure 8.2.

252 PRACTICAL ELECTRONICS

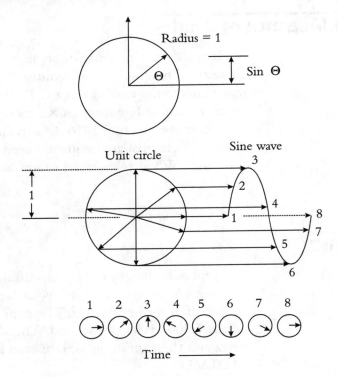

Figure 8.2 The sine wave function using a rotating radius

If the angle in Figure 8.2 increases with time, the value of h will move between + and −1 during every revolution. If the angle increases 360 degrees per second, then the height at any time t is equal to

$$h = \sin (360° \times t) \tag{8.18}$$

If the angle makes two revolutions per second, then the frequency f is 2 Hz The height h for any frequency f and for any value of time is

$$h = \sin (360° \times f \times t) \tag{8.19}$$

where f is frequency in Hz and t is time in seconds. For example, if f equals 10 Hz and $t = 0.025$ seconds, the angle is equal to 90° ($\pi/2$) and the value of h is 1.

We can use this sine function to represent any parameter that changes sinusoidally. For example, a sinusoidal voltage v is represented by the expression

$$v = V \sin(360° \times f \times t) \qquad (8.20)$$

The voltage v reaches a peak value of $+V$ once per cycle. If $f = 20$ Hz, then v varies between $+V$ and $-V$ 20 times per second.

Using radian measure for the angle, a sinusoidal voltage at a frequency f is given by the expression

$$v = V \sin(2\pi ft) \qquad (8.21)$$

where V is the peak value of voltage, f is the frequency in Hz, and t is time in seconds.

Rotating Pointers for Voltage and Current

Voltages or currents that vary sinusoidally are represented by a rotating pointer system. Each pointer rotates once per cycle. When a pointer points right, the voltage or current is 0. When that pointer is vertical, the current or voltage is maximum. The length of the pointers represents the peak value of the voltage or current.

It is easy to consider the pointers as rotating but being viewed by a strobe light. The strobe light is adjusted until one of the pointers points straight right. This pointer will be the reference pointer. The direction of the other pointers will show the timing relationships between various currents and voltages. If a pointer for current is vertical when a voltage pointer is horizontal, the current has reached its peak 90 electrical degrees ahead of the voltage. The angle between the voltage and the current is called a *phase angle*.

In circuits with resistors, capacitors, and inductors, the voltages and currents peak at different times. This is represented by pointers that are separated by phase angles.

The Current in a Capacitor

When a sinusoidal voltage is placed across a capacitor, the current that flows depends on how rapidly the voltage is changing. For a sine wave voltage, the voltage changes most rapidly at the zero crossing of voltage. The maximum rate of change is $2\pi f V_P$ where f is frequency and V_P is the peak voltage of the sine wave. The peak current that flows from Equation 8.12 is

$$I_P = C \times 2\pi f V_P \tag{8.22}$$

The current that flows is also a sine wave where I_P is the peak value. The ratio of peak voltage to peak current is called the *reactance* of the capacitor and is

$$X_C = V_P/I_P = 1/(2\pi f C) \tag{8.23}$$

where X_C is the reactance in ohms, C is capacitance in farads, and f is frequency in Hz.

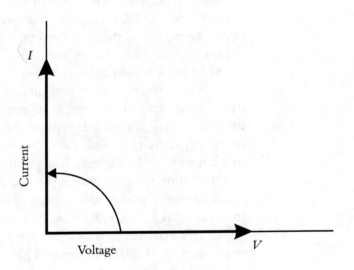

The pointers rotate once per cycle. This representation assumes the voltage is a sine wave. The current in a capacitor is also a sine wave. This 90-degree relation holds for all frequencies.

Figure 8.3 The rotating pointer system showing the timing relationship between current and voltage in a capacitor

The current in a capacitor for a sine wave voltage is always maximum when the voltage is 0. Figure 8.3 shows the timing relationship between current and voltage in a capacitor. Using the pointer system, you can see that the current leads the voltage by 90 electrical degrees. This relationship holds for all capacitors at all frequencies. Energy is stored in a capacitor when there is voltage. This means that peak energy is stored twice per cycle. Energy is stored in a capacitor; it is never converted to heat in a capacitor.

The Voltage across an Inductor

The voltage across an inductor is given by Equation 8.15. This voltage depends on the how rapidly the current is changing. If the current is sinusoidal, the maximum rate of change of current occurs when the current goes through 0. The maximum voltage occurs at this time. The resulting voltage is also a sinusoid, and the maximum voltage is

$$V_P = I_P \times 2\pi f L \qquad (8.24)$$

The ratio of peak voltage to peak current in an inductor is called *inductive reactance*. The equation for inductive reactance is

$$X_L = 2\pi f L \qquad (8.25)$$

where X_L is the reactance in ohms, L is the inductance in henries, and f is frequency in Hz.

Figure 8.4 uses the pointer system to show the timing relationship between voltage and current in an inductor. If the voltage is selected as the reference pointer, then the current pointer points straight down. This means that the current in an inductor lags the voltage by 90 electrical degrees. This relationship holds for all inductors at all frequencies. Energy is stored in an inductor twice per cycle when the current is at maximum. Energy is stored in an inductor; it is never lost as heat.

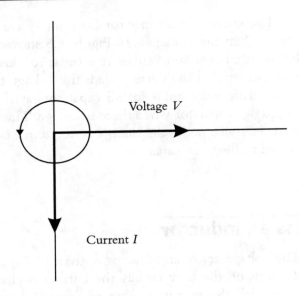

This representation assumes the voltage is a sine wave. The pointers rotate once per cycle. The current is a sine wave shifted 90 degrees at all frequencies.

Figure 8.4 The timing relationship between current and voltage in an inductor

The Meaning of AC and DC

The abbreviation dc stands for *direct current*. Direct current means that the voltage or current does not vary. A battery voltage is an example of dc voltage. If one ampere is drawn from the battery, this current is also called dc. Although the word "current" is present in the abbreviation, it is not always appropriate, as dc refers to both voltage and current.

The abbreviation ac stands for *alternating current*. The power voltages that we use are 60 Hz ac. Alternating current is a varying voltage or current. Again, the word "current" in the abbreviation may not be appropriate.

The Meaning of RMS

Voltages and currents are usually referred to in terms of their heating ability. At dc, the power dissipated in a resistor is $V^2(dc)/R$. It is very convenient if an ac source with the same measure provides the same heat. In other words, $V^2(ac)/R$ should provide the same heat. But when an ac voltage heats a resistor, the power varies during the cycle. The average squared value of a voltage is one-half the peak value. Thus we could say that power is equal to $V_P^2/2R$. Rather than involve this factor of two, the accepted practice is to refer to sine waves in terms of their peak value divided by the $\sqrt{2}$. This is their *rms,* or root mean square value. In this book all the ac voltages and currents are assumed to be rms values unless otherwise stated.

If V is the rms value of a voltage, the peak voltage is 1.414 V. The peak-to-peak value is 2.818 V. In our homes the rms value of voltage is 120 V. This is 169.7 V peak or 339 V peak-to-peak.

Square Waves

A square wave is a waveform where the voltage (current) transitions between two values. A typical symmetrical square wave might go from 10 V positive to 10 V negative at 10 kHz. Square waves do not have to be symmetrical; their average value may not be 0. For example, some square waves are at 0 V half the time.

Useful square waves must transition between voltages in a very short time. These transitions are called *rise and fall times.*

The rms value of a symmetric square wave is the peak value. If the square wave is 0 half the time, the heating value is one-half. This makes the rms voltage the peak value divided by the square root of two.

Square waves are a valuable tool for testing circuits. A square wave is made up of many sine waves. By using a square wave to test a circuit, the response can yield information not available by using a signal at one frequency.

The sine waves that make up a symmetrical square wave are called *harmonics.* The sine wave at the square wave repetition rate is called the *fundamental.* The next higher harmonic has a frequency three times the fundamental at one-third the amplitude. The harmonics are all odd multiples of the fundamental that have ever-decreasing amplitude. The next harmonic is five times the fundamental frequency at one-fifth the amplitude.

When a circuit is tested using a square wave, the response is to the fundamental and all the harmonics superposed. The peak amplitude of the fundamental sine wave is $2/\pi$ times the peak-to-peak value of the square wave.

Decibels

In electronics we often refer to ratios. The ratio might be of voltages at two different frequencies. It might be the voltage attenuation ratio in a filter. The ratios can sometimes involve very large or very small factors. The decibel provides a convenient way to handle these ratios. To understand the decibel, it is necessary to be familiar with logarithms.

The logarithm of a number is the exponent of 10 needed to equate to the number. $10^2 = 100$, so the logarithm of 100 is 2. The logarithm of 1,000 is 3, as $10^3 = 1,000$. Similarly, $10^{-2} = 0.01$, so the logarithm of 0.01 is -2. Logarithms squeeze the range of numbers between one-billionth and one billion into the range -9 to $+9$. The number to which the exponent is applied, in this case the number 10, is called the *base* for these logarithms. "The logarithm of 10^2" is written $\log 10^2 = 2$.

In the early days of the telephone, a unit was invented that represented the smallest detectable change in sound power level. This unit was called a decibel. The term bel was used to honor Alexander Graham Bell, the inventor of the telephone. The bel is the logarithm of the power ratio. The decibel was defined as $10 \log P_1/P_2$ where P_1 and P_2 are these power levels. When measured as voltages, this power ratio becomes $20 \log V_1/V_2$. For 1 decibel, the ratio of voltages is equal to 1.122. This assumes the voltages are applied to the same resistance. This logarithmic measure of ratios has come into common use, and the original power definition is sometimes lost. A circuit with a gain of 100 is said to have a gain of 40 decibels, abbreviated 40 dB. This measure has nothing to do with power.

In some areas of electronics, voltages are given in terms of dB volts. In this case the reference value or V_2 would be 1 V. 20 dB volts is simply 10 V. The same idea can be used for amperes or electric field strength. If the reference level is 1 mV, then 20 dB mV is the same as 10 mV.

The beauty of the decibel in describing performance is that decibels add while gain factors multiply. For example, if two circuits have a gain of 23 dB and 12 dB, the combined gain is 35 dB. If a circuit attenuates a signal by 45 dB and the gain following is 60 dB, the overall signal gain is 15 dB. If the initial signal was 5 dB volts, the resulting signal would be 20 dB volts or 10 V. The dB makes calculations very simple. See the table showing dB values.

Factor	dB value	Factor	dB value
1.0	0	1.0	0
1.414	3	0.707	−3
2	6	0.5	−6
3	10	0.3	−10
10	20	0.1	−20
100	40	0.01	−40
1,000	60	0.001	−60
10,000	80	0.0001	−80

Frequency Response

The response of a circuit to sine wave voltages depends on frequency. Usually a voltage amplitude is plotted on the vertical axis against frequency on the horizontal axis. Correctly speaking, this is an amplitude-versus-frequency response curve. It is often referred to simply as a frequency response curve.

The frequency axis is often logarithmic to accommodate a wide range of frequencies. Major scale divisions can represent factors of 10. The spacing between 1 and 10 Hz would equal the spacing between 100 kHz and 1 MHz. The vertical scale can represent gain, or it can represent relative amplitude. If the vertical scale is in dB, the vertical scale is already logarithmic. This is the preferred scale because many of the important features of the response can be easily seen. If a filter response falls off as the square of frequency, the plot of dB amplitude versus the logarithm of frequency is a straight line.

Appendix I
Preparing to Use the Learning Circuits

Components

The components you will need in the Learning Circuits are available in most electronics parts stores. Some stores carry loose stock, as opposed to individually packaged components. The advantage to loose stock is lower cost. Individually packaged items will be more clearly identified as to value and rating. As an example of cost, ½ W carbon resistors in bulk cost a manufacturer under two cents each. A single packaged resistor may cost 50 cents or more. This cost can add up if you need a hundred resistors.

The number of components you will need is not exact. You may want to leave some of the circuits assembled rather than tear them down for the next lesson, or you may want to experiment with constructions that are not a part of the Learning Circuits. This is entirely up to you. You may want to start slowly and buy a few parts. Buying the parts is part of the fun. It is also part of learning about electronics.

You will probably purchase resistors that are ½ W 20% carbon, although metal film resistors can be used. In many cases the exact values indicated are not critical and a nearby value will be acceptable. Except for size, 1-W resistors will work fine.

The smaller capacitors can be metalized Mylar, although other types will be quite adequate. The larger capacitors (above 10 µF) should be electrolytics rated 35 V or higher. In the voltage doubler circuit, the capacitor must be properly rated.

The power diodes 1N1002 can be any diode in the series 1N1001 through 1N1005. The last number defines the reverse voltage, and in our circuits this voltage is quite low. There are many other power diodes that would function in these circuits.

The transistor types suggested are readily available, although many other types would be acceptable. The best way to substitute another type is to look at specification sheets. There is often help available in most stores as to how you can make substitutions. Manufacturers provide component data on the Internet. You may want to try this approach. Just type in the part number on "Google" and you will be surprised. The only things that are important are the maximum voltage and current ratings and the gain, called β.

There are many ways to build experimental circuit boards. I have suggested an epoxy board with a grid of punched holes. Packages of pins are available that press-fit into these holes. Components can then be soldered to the pins. You might want to buy a package of 100 pins to get started. A tool to press-fit the pins into the board is often a help. The following list of components will allow you to build all the circuits in this book.

Resistors: ½ W 20% carbon except where noted

4 10 Ω, 100 Ω, 4 330 Ω, 4 470 Ω

6 1 kΩ, 2 2.2 kΩ, 2 3.3 kΩ, 2 3.9 kΩ, 2 4.7 kΩ

6 10 kΩ, 2 18 kΩ, 2 22 kΩ

2 100 kΩ, 2 1 MΩ

2 1.0 kΩ 1 W 20% carbon

Potentiometer

10-kΩ single turn carbon

Capacitors:

The majority of these components can be metalized Mylar rated greater than 50 V. Other dielectrics can be used. The voltage ratings should be 50 V or greater.

1 100 pF, 2 0.001 µF, 4 0.01 µF, 3 1.0 µF, 4 100 µF 35-V electrolytic

Semiconductors

2 1N1002 and 4 1N4148 diodes

2 10-V, 3 5.1-V, and 2 15-V zener diodes

1 red LED diode

2 TIP29A, 2 TIP30A, 3 2N3904, and 2 2N3906 transistors

1 LF353 IC amplifier

Inductor

10 mH. Under 10-Ω dc resistance. Natural frequency above 100 kHz.

Hardware

2 9-V and 1 1.5-V battery

tinned # 16 or # 18 10-foot bus wire

4 or more red and black clip leads of different lengths

1 SPDT switch

1 8-pin DIP socket for LF353 plastic IC

test board with a grid of holes approximately 6" by 8". An epoxy board is recommended.

1 package of 100 press-fit pins (must match the hole diameter in the board)

1 18-V ac adapter rated ½ ampere or greater; #PHC-AC-1888C or equal

soldering iron and stand

rosin core solder for electronics

needle-nose pliers

wire cutter

Soldering

You will need to know how to solder if you want to construct the Learning Circuits in this book. Soldering is the process of heating an alloy of tin and lead called solder and using it to make an electrical connection. It requires a supply of solder and a soldering iron to provide heat.

There are many types of soldering irons. The irons used for electronics are not the 100- or 200-W variety. These heavy irons can damage sensitive components through overheating. In electronics we use a smaller iron that has a small pointed tip, ideally one with a regulated temperature at the tip. 20 or 30 watts is adequate.

The solder you will use is usually provided on a spool. It looks like a coil of white uninsulated wire. In the center of the solder strand is a material called *rosin*. During the soldering process the rosin aids in cleaning the surfaces to allow the solder to flow.

Preparing the Soldering Iron

After your soldering iron is hot, make sure the tip can melt solder. If you have a new iron, melt solder on the entire tip area. Wipe off any excess solder with a heavy cloth. If you leave a soldering iron on for a period of time, a layer of oxidation or scum will form. Always wipe this layer of scum off the iron with a rag before you begin soldering. Heat cannot flow effectively through this scum.

The only way heat can properly flow from the soldering iron is through the melted solder on the tip. The tip is ready when it has a thin, bright white layer of melted solder on its surface. The heat from the soldering iron must flow to the conductors being soldered. Touch the soldering iron tip to the conductors to be soldered. Heat will flow to the conductors through the "tinned" tip. Feed the new solder by hand to the leads being soldered near the point of contact, not to the soldering iron. When the leads are hot, the solder will melt onto the leads.

What Is a Good Solder Connection?

A good solder connection is a thin layer of solder that connects two or more conductors. It is not just a glob of solder. The solder should flow so that the solder tapers to the lead. Any foldback of solder is suspicious. These poor junctions are called *rosin joints*. They can be a source of trouble. A rosin joint can be intermittent or even an open connection. This is a problem you do not need when you are getting started.

Soldering Hints

If you have done soldering in other applications, such as plumbing, you may have used acid flux to help with bonding. Acid flux is never used in electronic soldering. The acid can eventually do damage to components. The solder used in electronics contains a rosin that forms the core or center of the solder The rosin helps clean the surface and is adequate to do most circuit soldering.

One last tip: Avoid using a solder joint as a mechanical support, except on a very temporary basis. The right way is to use a mounting pin or terminal so that the connection is anchored to a circuit board. Solder is an electrical connection, not a mechanical support.

Obtaining Measuring Equipment

Ideally, you should buy or borrow an oscilloscope and a waveform (function) generator, so that you can see for yourself how circuits work. Both are readily available at any large electronics store. Unfortunately, they are not inexpensive—you may pay several thousand dollars for new ones. Used oscilloscopes and function generators are available for a few hundred dollars. When I checked on eBay recently there were many available. I know this is still a considerable investment, but if you get hooked on electronics you will need this measuring equipment and will use it over and over. Or ask around—perhaps someone you know has some equipment you can borrow.

However you obtain this equipment, you will learn a great deal from using it. There is truly no substitute for hands-on experience in electronics, and this is what I have tried to give you with the Learning Circuits. Have fun with them, and your understanding of electricity will grow very quickly.

Appendix II
Basic Algebra

Introduction

To read the text of this book and to work the problems, you need a basic understanding of the branch of mathematics called *algebra*. The tools of algebra enable us to do all the arithmetic operations without requiring us to use specific numbers. This generalization gives algebra tremendous flexibility and usefulness.

I have used algebra only where I felt it was really necessary, and this discussion is not meant to be a comprehensive explanation of algebra, but only a review of those topics I actually use.

If you have already studied algebra, this brief section will refresh your memory. If you have not studied algebra before, this section should give you a sufficient understanding to read the text. The only way to gain proficiency in using algebra is to work many problems. Unfortunately, there is not space in this book for those problems. If you want to really learn algebra, which I recommend, there are many textbooks available. A book in this series called *Practical Algebra* is recommended.

Symbols and Equations

The operations used in algebra are essentially the same ones used in arithmetic. The common operations are addition, subtraction, multiplication, and division. The symbol for addition is the + sign and the symbol for subtraction is the − sign. Simply placing two quantities next to each other indicates that they are to be multiplied. Sometimes the symbol × is used to indicate multiplication. In algebra, division is usually indicated by a horizontal line. (The division symbol ÷ used in arithmetic is not very convenient.) Items below the line are divided into items above the line.

The difference between arithmetic and algebra is that algebra uses symbols in addition to numbers. In arithmetic, you might write 2/3, which means "two divided by three." In algebra you can perform the same operation and write it the same way, but use the symbols a and b instead of the numbers. The term a/b means "a divided by b."

In algebra, the sequence of operations is important. The expression $c/(a + bc)$ means "multiply b times c, add the product to a, then take this sum and divide it into c." This is the only permitted sequence to this calculation.

Engineers and scientists often use symbols to represent parameters. Parameters are the things that vary in a problem. (They are also called *variables*.) For example, the letter d is often used to represent a distance. The distance might have units of feet, meters, or miles. When you drive your car, the distance you drive at a steady velocity for a given length of time is simply the velocity times time. The shorthand or algebra for this statement is $d = vt$.

The symbol = means "equals." In this example the letter v stands for velocity and the letter t stands for time. For example, if you drive 60 miles per hour for 2 hours the distance is 120 miles. In this case $v = 60$ miles per hour and $t = 2$ hours. In words, 120 miles equals 60 miles per hour times 2 hours. The statement that $v = d \times t$ or $v = dt$ is very compact and covers all velocities, distances, and periods of time for any system of units. Distance, velocity, and time are all called parameters in this statement.

The statements $v = 60$, $t = 2$, and $d = vt$ are known as *equations*. Equations are relationships between parameters, or between parameters and numbers. For an equation to work, the parameters must have compatible units. If the velocity is in miles per hour, the time must be in units of hours, not seconds. The units themselves can always be used in the form of an equation to see that everything is proper. For example, assume the velocity has units of miles/hour. This is read as "miles divided by hours" or "miles per hour." The equation using units for $d = vt$ reads

$$\text{miles} = (\text{miles/hour}) \times \text{hour} = \text{miles}$$

Notice that the unit "hour" appears in the numerator and denominator. Just as in arithmetic, you can cancel the two identical terms, leaving the unit miles on both sides of the equation. An equation is not valid unless the units agree on both sides of the equal sign.

The Number System

The simplest representation of the numbers we use places them on a straight line. The center point is 0. To the right are the increasing positive numbers and to the left are the increasing negative numbers. The addition of numbers simply adds lengths of the line from the origin. The length 3 is added to the length 5 for a value of 8. If both numbers are negative, the length is given a negative direction. The subtraction of numbers is the distance between points on the line. The distance between 6 and 8 is plus 2. This is written as $8 - 6 = 2$. The distance from 8 to 6 is a minus 2. This is written as $6 - 8 = -2$. The multiplication or division by a negative number reverses the direction of the original number. For example, minus 1 times a plus 6 is a minus 6. Minus 6 times minus 1 is a plus 6. A minus 6 divided by a minus 1 is also a plus 6. When symbols are used instead of numbers, they are all treated as if they were positive values. For example, $(-2)(-a)$ equals $+2a$. If a turns out to be a negative number, the answer will be negative.

About Parameters

When a parameter changes value, an equation relating those parameters must still hold true. The other parameters must adjust to maintain the equality. For example, if $d = vt$ and d changes by a factor of 4, then v and/or t must adjust. If v is fixed, then t must increase by 4. If t is fixed, then v must increase by 4 times. It is also possible for both v and t to increase by a factor of 2. Obviously there are many other ways for the product vt to increase by a factor of 4.

Algebraic Manipulations

Algebra allows changes to both sides of an equation so that the equality is maintained. By making these changes, new facts about the original relationship can be uncovered. Again, the rules of algebraic operation are the same as the rules of arithmetic. If $d = vt$, then algebra says that $4d = 4vt$. In the previous example, the distance was 120 miles. The new equation for the same parameter values states that 480 miles equals 480 miles. The rules are simple: There can be the same addition, subtraction, multiplication, or division to both sides of an equation. One restriction is that division by 0 is not permitted.

Changes to an equation that involve addition or subtraction must involve a separate term on both sides of the equation. Consider the equation $a/b = c/d$. Adding f to both sides is permitted. The new equation is $a/b + f = c/d + f$. Adding f in any other way is illegal. For example, $(a + f)/b$ is not equal to $(c + f)/d$.

Changes to an equation that involve multiplication must affect every term on both sides of the equation. Consider the equation $a + b = c + d$. If we multiply both sides of the equation by f, the result is $af + bf = cf + df$. All of the terms must be multiplied. The sum $af + b$ is not equal to $cf + d$ because only the first terms are included.

The expression a/b means "divide a by b." If $a = 10$ and $b = 5$, the result is 2. The expression a/b can be changed to $3a/3b$ without changing its value. The rule is very simple: The numerator and denominator of an expression can both be multiplied by the same terms and the expression remains unchanged. It is illegal to add or subtract the same term from the numerator and denominator. The expression $(a + 1)/(b + 1)$ is not equal to a/b. In multiplication, all the terms in the numerator and denominator must be included.

Equations remain unchanged if the reciprocal is taken of both sides. For example, if $a = b$, then we can divide both sides by ab. The result is $1/b = 1/a$. Of course, this can also be written as $1/a = 1/b$. This method is often used as a step in solving an equation for an unknown.

Algebra has some further rules that must be followed, or errors can result. The proper sequence of operations must be followed. The statement "three plus five times two" can be read two ways. The first way is to add three and five together and then multiply by two. The second way is to multiply five by two and then add three. The first answer is 16 and the second answer is 13. Algebra uses parentheses to indicate sequence.

If the first answer was intended, then the three and five must be placed in parentheses. The rules of algebra require that terms inside the parentheses must be treated as one value. In symbolic form, $(a + b)c$ is different from $a + bc$. Note that $(a + b)c$ can also be written as $ac + bc$. Of course, all of these rules can be checked using numbers.

Solving an Equation

Consider the equation $d = vt$. We can divide both sides of the equation by t and obtain the result $d/t = v$. We can do this by drawing a line under both sides of the equation. We then place t in the denominator on both sides of the equation. We do not know the value of d/t, so we leave it in symbolic form. On the other side of the equation, vt/t is the same as v because t/t is the same as 1. We say that the t's cancel. We can develop a third relationship by dividing both sides by v and obtain the equation $d/v = t$. These new equations are valid if the first equation is true. We could add or subtract any parameter from both sides of the equation without affecting the balance of values.

To see how the rules work, we can solve for R in the equation $(3R - S)/W = C$. First multiply both sides of the equation by W. The result is $3R - S = CW$. Next add S to both sides of the equation. The result is $3R = CW + S$. Dividing both sides by 3, we obtain $R = (CW + S)/3$. If the parentheses are removed from the original equation, it reads $3R - S/W = C$. We can add S/W to both sides and obtain $3R = C + S/W$. Now $R = C/3 + S/3W$, a very different expression.

Algebraic Identities

The following identities are always true and perfectly general. For example, the identity $a/a = 1$ can be extended to cover $b/b = 1$ or $(c + 1)/(c + 1) = 1$.

$$-(-1) = +1, \ +(+1) = +1, \ -(+1) = -1, \ +(-1) = -1$$

$$-(-a) = a, \ +(+a) = +a, \ -(a) = -a$$

$$a + b = b + a, \ a - b = -b + a$$

$$-(a + b) = -a - b, \ -(a - b) = -a + b$$

$$a/a = 1, \ -a/a = -1, \ a/(-a) = -1$$

$$ab = ba, \; a/1 = a$$

$$a(b + c) = ab + ac, \; (b + c)/a = b/a + c/a$$

Exponents

A notation that finds frequent use in algebra is the exponent. The simplest example of an exponent is the square. The expression a^2 is read "a squared" (or sometimes "a to the second power"), and it means $a \times a$. Similarly, a^3 ("a cubed") means $a \times a \times a$. Exponents are additive when the same base parameter is involved in multiplication. For example, $a^2 \times a^3 = a^5$. The expression $a^2 b^3$ does not equal $(ab)^5$. It follows that in division, exponents subtract. As an example, $a^5/a^2 = a^3$. A term to the exponent zero is one for all parameters. This is because $a^n/a^n = 1$ for any value of a or n, and $(n - n)$ is 0. The exponent 1 is redundant, as it leaves the value unchanged; $a^1 = a$.

The term a^2/a^5 can be written two ways. When the exponents are subtracted, the result can be a^{-3} or $1/a^3$. A simple rule follows: Any term can be moved from the numerator to the denominator by changing the sign of its exponent. The same rule applies when a term is moved from the denominator to the numerator. As an example, the term $a^2 b^{-3}/c^{-3} d$ could also be written as $c^3 d^{-1}/a^{-2} b^3$.

Fractional exponents can be used. The simplest application is the square root. The notation $a^{1/2}$ means the square root of a. This can also be written as $\sqrt{a}$. The definition of the square root is a number that multiplied by itself yields the number. The product $a^{1/2} a^{1/2} = a^1 = a$.

When an exponent is used with the number 10, the 10 represents the number of zeros following the one. For example, 10^2 equals a 1 followed by two zeros, or 100; 10^6 is 1,000,000; 10^{-6} is one-millionth, or 0.000001. In this case the exponent is the number of decimal places before the decimal point. The process of adding zeroes is referred to as powers of 10. Powers of 10 are much easier to use than large groups of zeros.

Adding or Subtracting Terms

Adding or subtracting terms algebraically is no different from adding or subtracting fractions. To add 2/3 to 1/7 we first find the common denominator. The common denominator is 7 times 3. Now we can change each fraction so that it has the same denominator. The two fractions become 14/21 and 3/21. The sum is (14 + 3)/21 or 17/21.

To add the terms a/b to c/d we can do the same thing. The common denominator is bd. To change a/b so that the denominator is bd, we multiply the numerator and denominator by d. The result is ad/bd. Similarly, the term c/d can be changed to cb/bd. With a common denominator, the sum can be written $(ad + cd)/bd = (a + c)/b$.

In chapter 2 the equation for parallel resistors was given as $1/R_3 = 1/R_1 + 1/R_2$. The righthand sum can be rewritten as $R_2/R_1R_2 + R_1/R_1R_2$. With a common denominator, the terms can be combined as $(R_1 + R_2)/R_1R_2$. Now the equation is $1/R_3 = (R_1 + R_2)/R_1R_2$. To solve for R_3, we simply invert both sides of the equation and $R_3 = R_1R_2/(R_1 + R_3)$.

Index

ac, 7, 251, 256
ac adapter, 3, 65
active current source, 127
active high-pass filter, 181
active low-pass filter, 178
A/D converter, 210
astable multivibrator, 204

base current, 96, 114–115
base/emitter junction, 95, 111
batteries,
    in parallel, 10
    in series, 9
beta, transistor, 97
bistable multivibrator, 201
bridge rectifier, 72

capacitance, 8
capacitors, 7, 23, 73, 245
    current flow, 24
carbon resistors, 15
centertap, 71
circuit board, 3
clock signal, 148
closed-loop gain, 153, 171
coaxial cable, 229
collector, 95
color code, 16
common conductor, 5

comparators, 194, 211
conductances, 12
constant current loop, 191
constant current source, 129, 191
crystal oscillator, 206
current divider, 21
current source, 126, 190

D/A converters, 209
D/A flash system, 211
dc, 7, 256
dc amplifier, 102, 116
dc power supplies, 65, 69
differential amplifier, 175
differential stage, 130
digital, 146
digital word, 207
diode clamp, 84
diodes, 65, 83
DIP socket, 154
direction of current flow, 83
distortion, 167–170
doping, 82
drain, 134

electric field, 218
electronics, 1
emitter, 95
emitter follower, 98, 103

error correction, 166
excitation voltage, 212
exponential curve, 49

farad, 44
feedback circuits, 152, 153, 166
ferrite, 63
FET. *See* field effect transistor
field coupling, 232
field effect transistor, 134–137
filter, active, 180
filter capacitors, 67
forward gain, 153
frequency, 251
frequency trap, 59
function generator, 4

gain, 102, 114, 162
gain at ac, 108
gate, 134
ground, 5
grounding, single point, 226

half-dipole antenna, 233
henries, 44
high-pass filter, 34, 53, 181
hysteresis, 200

IC amplifier, 156
impedance, 28, 50, 52
    RC circuit, 28
    RL circuit, 50
inductance, 44
induced voltage, 43
induction flux, 61
inductors, 41, 53, 247
input impedance, 100, 121, 165
integrated circuit amplifier, 154
integrated circuits, 151, 187
    linear, 152
integrator, 192
    digital, 194
interference, audio, 222

interference coupling, 227
internal resistance, 19

laminations, transformers, 62
Learning Circuits, 2
LED. *See* light emitting diode
Lenz's law, 43
light emitting diode, 137
load cells, 214
load resistor, 101
logic, digital, 147
low-pass filter, 30, 51
L/R time constant, 48

magnetic field, 218
magnetic material, 63
magnetizing current, 61
multimeter, 3
multivibrator, 201

National Electrical Code, 7
negative feedback, 152–153
npn transistor, 95

offset, 106
Ohm's law, 15
open-loop gain, 153
operational feedback, 163
oscilloscope, 5
output impedance, 101, 127

parallel resonant circuit, 59
pass element, 126
phase, 10, 171
phase angles, 30, 253
phase response, 34
phase shift, 172
phototransistor, 138
pn junction, 88
pnp emitter follower, 111
positive voltage regulator, 124
potentiometer, 115
potentiometric feedback, 157

Index

power supply
  rails, 113
  dc, 66
p–type transistor, 82

quartz crystal, 205

R-2R ladder, 208
radiation, 218, 235
RC low-pass filter, 30
RC time constant, 26
reactance, 46, 61
reference conductor, 222
resistance, internal, 19
resistors, 7, 11, 15, 16
  color code, 16
  metal film, 16
  parallel, 12
  series, 11
resonance, 55, 56
resonant frequency, 55, 59
ringing, square wave, 57
RL low-pass filter, 51
RLC circuit, 53
rms, 257
rotating pointers, 28, 50

saw-tooth waveform, 195
schematic, 71
Schmidt trigger, 200
SCR, 139–140
second-order filter, 179
semiconductors, 81–82
series resonance, 56
shield, 5, 220
signal controlled rectifier, 139
signal generator, 6
silicon, 82
sine waves, 4, 46, 251
single point grounding, 226
solenoid, 42
source, 134

source impedance, 90
square waves, 4, 34, 53, 257
stability, 171–173
stacked emitter follower, 120, 166
step function, 9
strain gauge, 212
summing point, 163
switches, 22, 142
switching power supply, 144

teslas, 63
Thevenin's theorem, 21
time constant, 28, 48
transconductance, 97
transformer, 61, 146
  secondaries, 64
transistor
  clamp, 197
  gain, 97
  switch, 138, 143
transmission lines, 227
  reflections, 230
triac, 140
trigger circuit, 199

virtual ground, 163
voltage attenuator, 115, 159
voltage divider, 17, 104, 121
voltage doubler circuit, 74
voltage gain, 106, 108, 116
voltage regulators, 91, 122, 188
voltage source, 7, 123, 126

wave, 228
wave form, 4
wave-form generator. *See* function generator
Wheatstone bridge, 212

zener diode, 88
zero reference conductor, 5